CLEARING THE AIR

First published in 2005 by Liberties Press
51 Stephens Road | Inchicore | Dublin 8 | Ireland
www.libertiespress.com | info@libertiespress.com
Editorial: +353 (1) 402 0805 | sean@libertiespress.com
Sales and marketing: +353 (1) 453 4363 | peter@libertiespress.com

Trade enquiries to CMD Distribution
55A Spruce Avenue | Stillorgan Industrial Park | Blackrock | County Dublin
Tel: +353 (1) 294 2560
Fax: +353 (1) 294 2564

ISBN 0-9545335-6-9

2 4 6 8 10 9 7 5 3 1

A CIP record for this title is available from the British Library

Cover design by Liam Furlong at space.ie
Set in 11.5 point Garamond

Printed in Ireland by Colour Books
Unit 105 | Baldoyle Industrial Estate | Dublin 13

CLEARING THE AIR
THE BATTLE OVER THE SMOKING BAN

NOEL GILMORE

DEDICATION

To the people who have made Ireland
a better and healthier place

Contents

ACKNOWLEDGEMENTS

This book is based almost entirely on information sourced from a vast number of media reports, articles and debates on the ground-breaking Irish smoking ban and from a wide range of publications on the topic of tobacco control. There are simply too many sources to mention, but each and every one deserves to be acknowledged in a general 'thank you'. I have made every effort to reflect these sources as accurately as possible and to use the information only in its proper context. If there are any inaccuracies, they are unintentional and hopefully will be seen in the light of this work being not so much an academic or in any way a pretentious exercise, but rather as a personally motivated attempt to mark what I regard as a major landmark in Irish social affairs.

A few specific acknowledgements are also necessary. Thanks to Jim Deere, who read a few early speculative chapters and said: 'Go ahead, I'd buy it.' To the key players who took the trouble to add comments or helped in a variety of ways. To Seán O'Keeffe and Peter O'Connell of the relatively new publishing company Liberties Press (which deserves every support), for their trust and help. To Jeffrey Wigand, a brave and inspiring man, for contributing the introduction. (Dr Wigand's role in revealing information about the tobacco industry was the subject of the film *The Insider,* in which Dr Wigand was portrayed by the leading screen actor Russell Crowe.) To NiQuitin CQ and GlaxoSmithKline, the maker of NiQuitin CQ, who have sponsored the distribution of *Clearing the Air* to health professionals and anti-tobacco agencies throughout Ireland. I would also like to acknowledge personally the excellent work of the Irish Cancer Society, not only in caring for cancer sufferers and their families, but also in tackling the causes of cancer, including smoking. Author royalties from the sale of this book are being donated to the Society.

Finally to Pauline, who, as ever, patiently supported me in another of my 'projects'. To Miriam and Niall for continuing to encourage their Dad, and particularly to my other daughter, Anita, for all her help with the word-processing, for her enthusiasm and for coming up with the title, 'Clearing the Air'.

INTRODUCTION

'CONGRATULATIONS TO IRELAND ON LEADING THE WAY'

Those of us who are committed to tackling the very serious world-wide problems caused by tobacco smoking are inspired by the courageous and ground-breaking stance taken by the Irish government in introducing smoke-free workplace legislation in March 2004. In particular, I would like to commend the lead role played by your minister for health, Micheál Martin, who showed such determination in carrying through his plan to make Ireland the first country in the world to eliminate tobacco smoke from the workplace. In doing so, he took a very positive and practical step towards protecting the health of many thousands of workers and members of the public.

I observed the tactics of the Irish anti-ban campaigners: the pattern of protest and objections was reminiscent of what invariably arises whenever anti-smoking (pro-public health/pro-clean air/harm reduction to innocents forced to breathe other's smoke) legislation is proposed anywhere in the United States, or in other countries. Minister Martin is to be congratulated on refusing to be intimidated in the face of the many challenges he encountered as he prepared to introduce the new law. His perseverance served to secure for Ireland a place in history as the country that set a new standard in tackling the evils of tobacco smoke.

There are more than 5 million tobacco-related preventable deaths in the world each year. This epidemic toll is expected to grow to more than a 'Holocaust' per year in the next decade unless the tobacco industry is held accountable for the products it markets to our children and is transformed through morally mandated regulations and the development of public knowledge on the issue. In the USA and Canada alone, more than 480,000 people – many of whom have never chosen to smoke – are dying each year from the effects of tobacco use and exposure to its lethal components. Tobacco's cumulative toll on humans exceeds that of many other causes of death, including AIDS, homicides, suicides, vehicular accidents and fire deaths combined. This toll is extracted in the form of lung cancer, heart disease, impotence, infertility and cognitive impairment in the unborn from the effects of second-hand smoke *in utero,* and in relation to prenatal and perinatal exposure. In the USA alone, tobacco-related diseases cost taxpayers more than $100 billion each year in direct medical costs and an additional $140 billion a year in lost workforce productivity.

During the last decade, some progress has been made to 'denormalize' tobacco, which has become an accepted part of our daily lives over more than two centuries. Progress has been made in characterizing the products for what they are: 'Products which, when used as intended, kill not only the user but also the innocent unborn to the innocent adult worker,' as I point out on my website (*www.jeffreywigand.com*). Progress has been made on many fronts, such as in labelling tobacco products, listing the additives they contain, removing 'healthy' monikers such as 'Mild' or 'Light', introducing price increases, requiring the product to have reduced ignition propensity, and creating smoke-free environments in countries, cites, states and municipalities around the world.

The tobacco industry continues to focus its $14 billion advertising and marketing efforts mainly on women, children, minorities, and those in developing countries. The industry's mantras of 'Hook 'em young, hook 'em for life' and 'We are in the nicotine-delivery business' remain their primary strategic goals, and the industry continues the amoral practice of generating obfuscation and controversy over the scientific and business issues relating to personal tobacco use and the creation of smoke-free environments. Moreover, despite the industry's claims that it has changed, 'this leopard still hunts the same way it has hunted for decades, for its behaviour and spots are still there,' as I state on the website.

Taking on the tobacco industry is never an easy task. I realized this fact soon after going to work for Brown & Williamson Tobacco Corporation – part of BAT Industries, the world's second-largest marketer of tobacco products – in 1989. My view was confirmed in 1995, when I decided to go public with what I had learned, seen and witnessed while working as the senior scientific officer in Brown & Williamson. The tobacco industry has far too many billions of dollars at stake to accept quietly any such challenges to its power. I have often said that, were it not for the help I received from health groups, the media, and some public representatives, Brown & Williamson would have litigated me to death.

The Irish minister for health and his officials and tobacco-control allies faced many of the same challenges that I encountered as the highest-ranking former tobacco company executive to address the public health and safety issues involving tobacco products. The hospitality industry, with the direct and indirect backing of the tobacco industry, invariably warns of economic disaster following smoke-free legislation being introduced. I am sure the economic disaster never followed in Ireland, just as it has not followed in any other jurisdiction that has had the courage and foresight to pass similar laws. The government and health alliance faced up to the opposition, and in spite of the obstacles that stood in its path, they held true to their duty and convictions, and in the end they won the battle.

This book provides a very interesting and detailed insight into the battle for hearts and minds that was fought in Ireland over the 'smoking ban'. As such, it is an ideal case study that deals with the issues, problems and developments that were encountered as the process of preparing for the new legislation progressed. It highlights the intense media and public interest in the debate on smoking in the workplace and clearly illustrates that the anti-ban lobby was prepared to go to great lengths to put pressure on the minister for health and his government colleagues. It also sets out in detail the very cohesive approach pursued by the alliance of health interests that won an impressive level of public support for the new law.

Clearing the Air contains a number of lessons for those who are planning similar initiatives to the Irish ban. Some of the experiences they will face may not be as colourful as a number of the anecdotes in this book, but the communications obstacles will undoubtedly be very similar. In particular, I regard this book as being a means of complementing the efforts of everybody who supported the concept of introducing the smoking ban. It is the author's way of saying 'well done' to all concerned – and I would like to join with him in saying just that.

Today, Ireland stands as a stellar example of what can happen when an enlightened leader and his supporters have the vision to protect innocent citizens from the harm that is foisted on them by breathing another person's smoke. Elected leaders and supporters of the right to clean air throughout the world have a template of 'doing the right thing' and have no excuse for not following the course set by Ireland, given the preponderance of the scientific evidence relating to the harm caused by tobacco smoke to innocent people. The people of Ireland overwhelmingly agree with this new public-health action.

The Irish government and its supporters followed the moral imperative stated by John Stuart Mill in 1864 in his essay 'On Liberty':

> that the only purpose for which a power can be rightfully exercised over a member of a civilized community, against his will, is to prevent harm to others.

I leave you with this final thought:

> Thou shalt not be a victim.
> Thou shalt not be a perpetrator.
> Above all, thou shalt not be a bystander.

J. S. Wigand, Ph.D., MAT
SMOKE-FREE KIDZ, INC.
January 2005

PROLOGUE

'Funny place – I can't see the air' quipped the late Bob Hope when he stepped from a light aircraft on a golfing visit to the remote Atlantic-seaboard village of Waterville in County Kerry some years ago. Later that day, as guest of honour at a traditional 'Irish social', Hope could certainly see the air in the typical smoky atmosphere of the Skelligs Bar. Cigarette smoke, including second-hand smoke, is the direct cause of upwards of seven thousand deaths in Ireland each year. In fact, 90 percent of lung cancers are caused by smoking, 50 percent of smokers will die from smoking-related diseases – with smoking also being a major cause of heart disease, Europe's biggest killer – and it is known that these diseases are the most preventable of all causes of death.

Irish people and visitors to our shores no longer 'see the air' in Irish pubs as a result of one of the most outstanding – some would say courageous, others controversial – political decisions ever taken in Ireland. On 29 March 2004, legislation prohibiting tobacco smoking in any place of work was introduced by the then Minister for Health and Children, Micheál Martin TD. The implementation of the new law marked the conclusion of one of the most vehemently debated issues ever in Irish social life. The arrival of a 'smoke-free at work' environment throughout the Republic of Ireland, which was warmly welcomed by health-promotion interests and aggressively challenged by the hospitality sector and especially by the licensed trade and Big Tobacco, was truly an historic landmark. It was also a very important development from an international perspective, because Ireland was seen as a test case for such legislation. No other sovereign nation, as a whole, had introduced a total ban on smoking in the workplace. So implementation of the ban in Ireland was seen by tobacco interests as the 'thin end of the wedge'. Equally, the eyes of anti-smoking campaigners internationally were firmly fixed on the outcome of the Irish initiative. If it was successful, it would act as an important precedent worldwide.

This is an account of much of the comment, discussion and argument – very often heated and hostile – that was vented both for and against this initiative. It was a fascinating time for those with an interest in communications. The final outcome was certainly remarkable: it marked the conclusion of an important chapter in Irish politics, public affairs and, notably, media involvement in an important issue of public policy.

The events and commentary over the months leading up to the introduction of the ban provide food for thought on the potential for other initiatives which could affect public behaviour and society in general. Some examples are the potential for tougher laws and enforcement to deal with problems such as alcohol abuse and dangerous driving and creative legislation to tackle environmental protection, traffic management, obesity and other issues. Most importantly, this account sets the scene for the possibility that the legislation introduced in Ireland may over years to come be regarded as the benchmark for similar anti-tobacco initiatives worldwide.

1

'A Complete, Total and Absolute Ban on Smoking in the Workplace'

'Turkeys voting for Christmas' is how Frank Fell, chief executive of the Licensed Vintners Association, described bar workers who would support the ban on smoking in the workplace. He was speaking in January 2003, hot on the heels of the announcement of proposed new legislation due to be introduced by the Irish government the following January. Fell, representing the organisation that works on behalf of publicans in the greater Dublin area, stated that the ban would have 'no discernible benefit, only economic disadvantages'. His views were reinforced by his counterpart in the Vintners Federation of Ireland, Tadg O'Sullivan, who issued dire warnings of job losses in the pub trade throughout Ireland following implementation of any smoking ban. 'A total ban would be unworkable, untenable and unenforceable, and ventilation systems are effective in extracting smoke' was the immediate reaction of the VFI, a comment that was to constitute the central platform of their anti-ban campaign over the subsequent months.

The Irish Hotels Federation stated that it could not be expected to enforce a ban on smoking in Irish hotels. Hoteliers felt that it would be unreasonable for them to be expected to police a no-smoking policy throughout their premises. The arguments they put forward included the ineffectiveness of smoke alarms in deterring guests from smoking at open windows and the likelihood that guests who did not heed the ban could check out of the hotel before being discovered. The first shots were being fired in what would prove to be an intense, sometimes hostile, always contentious debate on the banning of smoking in the workplace in Ireland.

On 30 January 2003, then Minister for Health and Children Micheál Martin TD had announced: 'I'm banning smoking in the workplace, including restaurants, trains and pubs.' Speaking in a very forthright manner at a media briefing, the minister made it clear that he was taking very seriously the findings of a comprehensive study of reputable international research aimed at identifying 'the degree of consensus that exists among leading international scientific authorities on the question of the hazard and risk posed by environmental tobacco smoke to human health in the workplace.' The study was conducted by a team of independent scientists, chaired by Dr Shane Allwright. Dr Allwright, senior lecturer in epidemiology in the Department of Public Health and Primary Care at Trinity College Dublin, is widely respected for her work in the areas of respiratory health, passive smoking, alcohol misuse and blood-borne viral infections. All of those involved in the study, which was commissioned by the Office of Tobacco Control and the Health and Safety Authority, had signed a declaration that they had no vested interest in the tobacco industry, in smoking-cessation products or in any anti-tobacco lobby group.

The findings of the study challenged the view promoted worldwide by the tobacco industry, which had always robustly defended its products, arguing that there was no conclusive proof that passive smoking is harmful to health. The Irish expert team, having studied 126 separate academic papers and studies by researchers in the US, Australia, Canada, New Zealand, Japan, France, Luxembourg, Belgium, Switzerland, Hong Kong and Ireland, and including work published by the World Health Organisation and many other authorities worldwide, concluded, without equivocation, that environmental tobacco smoke is a cancer-causing agent and as such was a major contributing factor to the more than 7,000 cancer-related deaths in Ireland annually.

It did not take long for the ground-breaking significance of the announcement of the minister for health's initiative to sink in and for vested interests to sit up and take notice. He was bringing forward legislation which was not only new to Irish law but which constituted a ban which had no precedent in any other country in the world. Although complete or partial bans on tobacco use were in existence in a substantial number of individual states in the US and in particular regions elsewhere in the world, most notably in many municipalities in Canada and Australia, parts of New Zealand, parts of Asia and Africa, and a number of European states, no national legislation existed in any other country aimed at curbing the harmful effects of tobacco smoke to such a degree as the Irish proposed legislation. It was Ireland versus Big Tobacco.

Initial reaction to the announcement was certainly one of surprise. It amounted to little more than an initial skirmish or flexing of the muscles by the main interest groups that would be affected by such a ban. This was partly due to the interest groups being taken somewhat off-guard and having insufficient notice to prepare immediate, considered responses – and also due to the fact that the minister was giving almost a full year's advance notice of the implementation of the ban. There was a feeling among the anti-ban elements that there was plenty of time for preparations for a full debate and for efforts to persuade the minister to change his mind.

In truth, in the week the proposal for the ban was announced there was more media interest in the announcement of Brian Kerr's appointment to the job of manager of the Irish soccer team. Would he manage to entice Roy Keane back into the Irish squad after the notorious events of the World Cup in Saipan? Pages of newsprint and hours of broadcast airtime were allocated to this story. By contrast, initial coverage of the 'smoking ban' story fizzled out within twenty-four hours of the announcement.

It didn't take long, however, for the potential impact of the announcement to start to register in the pub trade and amongst the public at large. The 'smoking ban' debate came centre-stage in Irish bars as pub owners started to realise the implications of a ban for their businesses and began to generate discussion on the subject among their patrons. After all, the pub was very much the focal point of local politics: many political meetings took place there, and it was the hub of conversations on all matters political. Moreover, the pub trade had long been amongst the strongest lobby groups in the country and had a reputation for blocking mooted legislative changes, especially in relation to amending licensing laws to allow for deregulation, changing opening hours, and so forth.

The ban was soon to become the most contentious issue of conversation in pubs and workplaces throughout the land. Family members argued about the merits and pitfalls of the proposal, much as they did about the rights and wrongs of the 'Saipan affair'. 'Letters to the Editor' on the subject began to appear in the national press. A wide range of views were expressed in public and through the media by the business community, health professionals and ordinary smokers and non-smokers alike.

Disbelief is probably the most accurate description of the initial public reaction and attitude to the announcement. The planned ban was so radical – and so unexpected – that many people found it difficult to accept that the proposal was genuine. There was even a level of disbelief amongst some officials in the minister's own Department. There was certainly considerable disbelief in media circles. In most circumstances, new legislation is flagged well in advance. Indeed, very often legislation is introduced as a result of media-fuelled public debate on matters of concern to the population as a whole. As the essence of politics is to legislate for the common good, inevitably the public at large will have had their say, or will have had the benefit of arguments for and against being thrashed out in the media in advance of new laws being introduced. While there had been growing awareness over many years of the increasing incidence of cancer-related deaths and illnesses, however, at no stage had the impact of environmental tobacco smoke in the workplace either dominated the headlines or attracted any particular interest from 'the man on the street'.

Whatever about workplaces in general, the idea that, under the terms of the ban, there would be no smoking in Irish pubs, which were such a visible and traditional element of Irish culture, was unthinkable to the vast majority of the general public. Throughout Ireland, the pub was one of the major urban tourism attractions, while in rural areas it was the centre of all social activity: the leisure centre, the community meeting place, and the venue for celebrations ranging from christenings to family milestones, funerals, musical entertainment and weddings. An Irish pub with no tobacco smoke? Surely somebody must be joking.

This was the theme which not only the Irish but also the international news media latched on to. In the UK, BBC News reported on the massive backlash for which the Irish government was bracing itself as a result of the announcement. Sky News covered the story extensively, as did all the international news agencies, and it was a major news item on Australian TV.

Also attracting widespread interest, especially amongst the political pundits, was the impact that possible implementation of the ban would have on the political career of Micheál Martin. The success or otherwise of his proposed ban on smoking was interpreted as a test of the political strength of a man who was often viewed as a possible successor to Taoiseach Bertie Ahern, and especially of his strength within his own Fianna Fáil Party.

Martin, a popular former minister for education, had inherited the health brief from one of the other Fianna Fáil front-runners for leadership of the party, Brian Cowen. Cowen is famously said to have described his time in the Department of Health as like 'serving in Angola', due to the many political minefields he encountered during his tenure there.

Martin had set about introducing widespread reforms to the health services, but he, like his predecessors Cowen and Brendan Howlin, a minister in the previous 'Rainbow Coalition', was finding the battle for resources and the implementation of new strategies to be an Everest to climb. He would later be accused by the anti-smoking ban lobby of using the ban as a smokescreen to distract attention from the problems he experienced introducing reform into Irish health services.

Some observers wondered about just how committed the minister was to carrying through this new legislation. After all, the ban was to come into effect on 1 January 2004, almost a full year away: there was ample time for a change of heart.

Some of the mainstream radio chat shows treated the announcement lightly. Marian Finucane, on her top-rated programme on RTÉ Radio 1, tauntingly questioned how the minister's 'tobacco police' would be received in Irish pubs by New Year's Eve revellers on the first day of the ban. This was a theme Finucane would persist with on innumerable occasions. Smokers also phoned Joe Duffy's *LiveLine* show expressing disbelief at the very notion of not being allowed the little luxury of puffing a cigarette with their pint of Guinness. 'Daft', 'crazy', 'unbelievable' and 'out of touch with reality' were some of the milder expressions used live on the airwaves.

So what was to lie in store for Micheál Martin in the weeks and months ahead? Was his political future, already perceived by many media commentators to be in the balance, to be sealed by the outcome of the smoking ban? Would he survive the reaction to the ban from Fianna Fáil backbenchers, and from his fellow Cabinet members? Would he successfully defend his policy against the highly combative reaction of the industry sector, whose fate, it was claimed, would be sealed by the new law?

These questions would be answered in due course, but one question appeared to be answered on day one. Was Micheál Martin staking his reputation on his commitment to an unprecedented health initiative with his plan? The answer appeared to be an unequivocal 'Yes'.

'I believe that in every decade we are presented with one major choice; a choice where, if we call it right, we change the future for the better' was Martin's rallying call at the announcement of plans for a smoking ban. He was openly embracing the recommendations of the expert group and their clear understanding that passive smoking was a cause of both cancer and heart disease. He also stated his concerns about the impact of passive smoking on children in particular. 'We either give them the clean air to which they are entitled, or we force them to breathe the fifty carcinogens and disease-causing chemicals in secondary tobacco smoke,' he said.

Of particular note was Martin's insistence that he was heeding the expert advice that a partial or selective ban on smoking simply would not work. His plan was clearly for an all-out ban. As a result of this approach, he would be facing into a tough period of confrontation, especially with the Irish hospitality industry, which was preparing to urge compromise and the watering down of the pro-posed legislation.

On 30 January 2003, Minister Martin stated that he was intro-ducing a 'complete, total and absolute ban on smoking in the work-place'. The die was cast. Time would tell if he had the determina-tion – and, according to many, the political guts – to pull off some-thing which would focus worldwide attention on Ireland and in par-ticular on the traditional image of the Irish pub.

2

Preparing to Win Hearts and Minds

The smoking ban would not dominate the national and local news headlines again until the beginning of summer 2003, but in the interim period the behind-the-scenes activity on the issue was intense. Both pro-ban and anti-ban campaigners busied themselves in preparation for the prolonged battle that was about to get under way to win the hearts and minds of the general public and key opinion-formers, including politicians, employers, commentators and all those who would have a role in implementing the ban. It was not difficult to see, from their initial public comments in response to the announcement, the positions that would be taken by all the main players, for and against the ban.

Those in support of the ban knew from international experience that the challenges ahead would be robust and difficult to overcome. They knew that the anti-ban lobby would immediately start to work on the arguments against the anti-smoking initiative. They were also aware that the 'anti' campaigners would be expected to be very well resourced in terms of research, funding and expert assistance.

The core issue on which all interest groups would focus in the initial stages was the totality of the ban. Dr Shane Allwright and the other members of the expert group that had advised the minister, including Dr James McLaughlin of University College Dublin, Dr Dan Murphy of the Health and Safety Authority, Dr Iona Pratt of the Food Safety Authority of Ireland, Professor Michael Ryan of UCD and Dr Alan Smith of the Royal College of Physicians in Ireland were unequivocal. 'In the light of the adverse health effects of environmental tobacco smoke and the ineffectiveness of ventilation in removing the risk posed to health, the scientists state that legislative measures are required to protect workers', the report concluded. On the advice of the group, only a total ban on smoking in all workplaces would be effective in protecting employees, who were regarded as being exposed to a serious health hazard. The experts emphasised that partial bans did not work. They concluded that exposure to second-hand smoke was an infringement of the basic human right to good-quality air, that in workplaces where smoking was permitted employees were exposed to a known risk of cancer which could not be dealt with adequately by means of ventilation and that, therefore, the only practical means of protecting employees was through legislation prohibiting smoking in workplaces. In other words, there was no room for compromise on health grounds. The minister had been advised to go for a total ban.

The pro-ban lobby was mainly comprised of the minister and his officials in the Department of Health and Children, as well as the Department's own Health Promotion Unit, the Office of Tobacco Control and the Health and Safety Authority. A number of influential non-government organisations also lined up firmly behind the minister. These included the Irish Cancer Society, the Irish Heart Foundation, the Irish Medical Organisation, the Irish College of General Practitioners, the Asthma Society, and in particular the anti-tobacco campaigning organisation ASH (Action on Smoking and Health) Ireland. Of critical significance was the support of the trade unions. The three largest unions in Ireland – Mandate, which represents the majority of unionised workers in Irish bars, SIPTU, which represents hotel workers, and IMPACT – with a combined representation of 310,000 workers, voiced their support for a ban.

On the anti-ban side, the initial objections were raised by the Licensed Vintners Association, representing the Dublin publicans, the Vintners Federation of Ireland, representing publicans nationwide, the Irish Hotels Federation, the Irish Cigarette Machine Operators Association, and to a lesser extent IBEC, the Irish Business and Employers Confederation, which, it was claimed, seemed to be drawn into the anti-ban argument at the behest of business interests, including the tobacco, hotel and drinks industries. It was also significant that many of the top legal, research and issues-management consultancies in Ireland were contracted to tobacco interests. If legislation banning smoking in the workplace could be introduced in Ireland, it might represent the 'thin end of the wedge' in terms of ant-tobacco legislation internationally. Therefore, no expense would be spared by the drinks and tobacco industries in drafting in heavyweight support to oppose the ban.

Another very vocal opponent of the ban was the Irish Hospitality Industry Association. The IHIA, which ostensibly represented the pub, restaurant and hotel industries, was set up specifically to challenge the ban. The establishment of this organisation gave rise to the suspicion that tobacco interests were being represented in a new guise. After all, with publicans throughout the country represented by the long-established LVA, and VFI and hoteliers represented by the influential IHF, what need was there for a new organisation?

Of all the representatives of hospitality interests, Jim Murphy, president of the Irish Hotels Federation, was the least hostile to the ban in the early stages after the announcement of plans for it. His concern for hoteliers centred on how to handle customers who might refuse to abide by the ban. While the minister had decided in favour of a total ban, he needed to understand the predicament of hotel and guesthouse owners who would feel inadequate in enforcing the new measure. After all, they would find it difficult to police what patrons did in the confines of their hotel bedrooms.

The pro-ban lobby had some very strong evidence to support their case. This evidence had been highlighted in the research report submitted to Micheál Martin and subsequently adopted by him as the basis for his legislation. This evidence would now be translated into powerful, hard-hitting messages:

§   Some seven thousand deaths occur in Ireland each year as a result of smoking-related illnesses.
§   Exposure to passive smoking increases the risk of stroke by 82 percent.
§   The risk of lung cancer is increased by 40 percent as a result of passive smoking in the workplace.

§ There is no safe level of exposure to environmental tobacco, and ventilation systems do not provide adequate protection from its effects.

The Office of Tobacco Control (OTC), which had been established by Minister Martin with a view to overseeing the implementation of a national tobacco-control programme, and the Health and Safety Authority (HSA), which holds responsibility for safety in the workplace, had combined their efforts in commissioning the report. Tom Power, chief executive of the OTC and a former senior official in the Department of Health, is credited by Minister Martin as being the main driving force within his Department in identifying the need for new legislation and persuading Martin to give a high priority to tackling the passive-smoking issue. 'Tom Power had a clear view on the tobacco industry and the impact of tobacco on society,' said Martin. 'It was a classic situation where a minister meets an official and they are both on the same wavelength. When that happens, things can happen in legislation, beyond what people tend to expect.'

At the same time, Tom Beegan, chief executive of the HSA, was pressing for legislation to highlight and deal with the dangers of passive smoking in the context of worker protection. The HSA would subsequently play a key influencing role in the overall campaign for a ban on smoking in the workplace by declaring secondary smoke to be a known carcinogen and, consequently, a workplace hazard. This would have a major impact in highlighting the responsibility – and indeed potential liability – of employers in relation to this issue.

The Irish Cancer Society (ICS), through its chief executive, John McCormack, and smoking-cessation expert Norma Cronin, welcomed the total nature of the ban as a major landmark, but stressed that more efforts should also be made to assist people to give up smoking. The ICS would play a major part in persuading employers and the public at large to understand the need for the new law. Their ongoing anti-smoking campaigns would also be of vital importance in putting the ban in context, and their smoking-cessation programmes would be invaluable when the time for enforcement arrived. In particular, Norma Cronin, who as health-promotion manager for tobacco control at the ICS had played a major part in establishing the 'National Smokers Quitline', now found herself in the pivotal position of coordinating an alliance of health-promotion interests.

Predictably, Professor Luke Clancy and Dr Fenton Howell, noted long-term anti-smoking campaigners, representing ASH Ireland, voiced strong support for the ban. They stressed that, if the ban was to have a real impact on the health of workers, enforcement would be crucial. In addition, it would be necessary to provide a wide-ranging educational campaign, directed towards workers, on the effects of passive smoking and workers' entitlement to protection from these effects. Both Professor Clancy, a respiratory specialist, and Dr Howell would lend their voices as major champions of the ban in the ensuing months. Clancy announced that he was especially encouraged by the comment from the minister that 'There is no moral option open to me other than to take action.' The specialist applauded this clear resolve, which ASH Ireland regarded as a positive response to the organisation's ongoing efforts to secure tough legislative interventions to combat tobacco smoking.

The president of the Irish Heart Foundation, Dr Brian Maurer, threw his support strongly behind the scientific evidence that environmental tobacco smoke (ETS) is harmful and stated that the evidence had been endorsed by all major independent scientific bodies. Although much emphasis had been placed on the fact that ETS was a cause of lung and other cancers, the experts were in no doubt that ETS was also a cause of heart disease and other respiratory problems in adults and children.

A tripartite body comprised of the Irish Heart Foundation, the Irish Cancer Society and ASH Ireland was quickly formed. This body was to provide unified expert support for the pro-ban stance as the communications battle got under way.

Support for the ban from the trade unions initially came in the form of a comment from John Douglas, deputy general secretary of the bar workers' union Mandate, who announced that research had shown passive smoking to be the cause of death of 150 bar workers each year. Douglas had been making representations about the risks of passive smoking with regard to employees for a number of years, on the grounds of health and safety, but had found that the absence of international precedent had been an obstacle to change being sought. Interest in the topic among the unions was rekindled at a conference in Dublin, attended by Douglas, where the hazards of passive smoking were discussed. 'The argument for a partial ban was totally discredited at the conference and the upshot was that the three employee organisations, Mandate, SIPTU and Impact, collectively agreed that it would be in the best interests of our members to have an all-out ban,' Douglas recalls. 'Bar workers in particular needed to be protected because they were directly exposed to a greater extent than virtually any other employees because of the smoky environment they worked in over extended periods.'

With almost 70 percent of the adult population in Ireland being non-smokers, the opinion of the public at large in relation to passive smoking was interesting. Research conducted by pollsters TNS/MRBI indicated a surprisingly high level of support for banning smoking in various public settings. While substantial majorities, somewhat predictably, existed in favour of smoke-free shops, crèches, waiting rooms, cinemas, schools, universities and hairdressing salons, the level of support for smoke-free bars and restaurants was remarkably high: 67 percent of people stated that they supported the proposed ban, with 77 percent of non-smokers and 37 percent of smokers in favour of it. The research showed that support for the ban was evenly spread between men and women and among people of all ages: 69 percent of students and 76 percent of people over fifty years of age supported it.

From the outset, the positions of the main players in this intriguing battle for the hearts and minds of the Irish people were clear. Undoubtedly, an intense and possibly bitter struggle lay ahead; at this early stage in the campaign, it was not possible to say who would eventually emerge as the winner.

*

One of the most intriguing aspects of the Irish announcement of the ban was the fact that it would be setting what many regarded as a worldwide precedent. This was the cause of a great deal of comment, especially in relation to Micheál Martin's motivation in announcing the ban. It was also the reason why the major international tobacco interests would focus so strongly on the outcome of the initiative: although the legislation would come into force in a market that consisted of only 3 million adults, the Irish experience had the potential to serve as an international test case.

Ireland, along with 191 other countries, was a party to a legally binding treaty under the auspices of the World Health Organisation (WHO) called the Framework Convention on Tobacco Control. This treaty represents unanimous political consensus that passive smoking causes diseases, and it obliges governments to provide protection to third parties from environmental tobacco smoke. Signatories to the treaty are committed to 'adopt and implement effective measures providing for protection from exposure to tobacco smoke in indoor workplaces, public transport, indoor public places and, as appropriate, other public places in areas of existing national jurisdiction.'

Irish EU Commissioner David Byrne arranged for the Commission to adopt a proposal in June 2002 recommending a call to all member states to 'provide adequate protection from exposure to passive smoking at the workplaces, in enclosed public places and in public transport and to strengthen smoking-prevention programmes.' This was the most significant initiative to date involving European countries collectively in combating environmental tobacco smoke. The proposal was enthusiastically supported by the Irish health minister but was not readily adopted across all member states: both the commissioner and the minister had to exert pressure on Germany in order to reach a common agreement on tobacco.

As a party to the WHO treaty and the EU initiative, Ireland was adopting a lead position in the fight against tobacco interests; it found itself well ahead of other EU member states in approaching the elimination of smoking in the workplace. Elsewhere in Europe, the other countries planning, in the longer term, to introduce smoke-free legislation included Norway, the Netherlands and Sweden, while in Germany, France and the UK – and particularly Scotland and Northern Ireland – the smoke-free issue was gaining political momentum. Even Greece, the nation with the highest rates of smoking in the EU, was in the process of preparing to ban smoking in public places, as was Spain, the EU's second-largest consumer of cigarettes per head. The Italians, who had operated a ban on smoking in public buildings since 1995, were also considering a workplace ban, as were the Portuguese. Many other countries had partial bans, though some were not always enforced.

In the United States, the link between passive smoking and various diseases, including lung cancer, had first been identified by the Surgeon General in 1986. From that time, there had been growing acceptance in the international scientific community of the harmful effects of passive smoking. As evidence of the linkage grew, so did increased public awareness of the issue, leading to a number of successful court cases in the United States. One of the most significant of these cases was a class action brought by flight attendants in 1997, which resulted in a settlement of $300 million. California was the first state to introduce smoke-free-workplace legislation. The ban, which started with the elimination of smoking in bars in 1995, was extended to restaurants in 1998. Other states where total bans were later enforced included Connecticut, Delaware, Maine, Massachusetts, Rhode Island and New York. There were also smoke-free initiatives in many other places, including Los Angeles and San Francisco; in Georgia, smoking in a vehicle in the presence of a child under four years was banned.

The legal implications of the risks to employees exposed to ETS continued to gain momentum. In Australia, for example, a barmaid was awarded $235,000 in compensation for having contracted cancer as a result of working in a smoky environment for eleven years. The Netherlands was also to join the list of countries where employees were moving towards suing their employers on the grounds of health damage resulting from environmental tobacco smoke.

In Canada, a number of provinces and municipalities had enacted smoke-free laws. In the province of British Columbia, however, legislators were found not to have properly consulted with the parties involved, and the proposed new law was struck down by the province's Supreme Court. New Zealand was also moving towards smoke-free workplaces at the time the Irish ban became law; it introduced its own ban in December 2004.

But a total ban? No country had an outright ban on smoking in the workplace. The extent of the laws and legal restrictions varied throughout many countries worldwide; only in individual states and regions, rather than in entire countries, had total bans been enforced.

The decision in Ireland to introduce a total ban was not taken overnight. Apart from the previously mentioned Tom Power of the Office of Tobacco Control and Tom Beegan of the Health and Safety Authority, Tom Mooney, an assistant secretary at the Department of Health and Children, and a number of his colleagues are credited with consistently highlighting the harmful effects of environmental tobacco smoke. The input of Dr Michael Boland, chairman of the Office of Tobacco Control, was another key factor in the minister's decision. As president of WONCA, the World Organisation of Family Doctors, Dr Boland had already established a strong reputation internationally for introducing innovative programmes to tackle the smoking problem. He was involved in projects to encourage GPs to help their patients quit smoking and to eliminate smoking from all environments associated with health care, including GP waiting rooms, hospitals and other public health locations. Dr Boland was also involved in pressing for government interventions on this issue, including the imposition of high taxes on tobacco products.

The ultimate motivating factor for the minister was the collective analysis of the independent, multi-disciplinary team of experts that had produced the Health and Safety Authority/Office of Tobacco Control report. 'This report gave categoric advice both to myself and the Minister for the Environment that environmental tobacco smoke causes cancer,' Micheál Martin commented.

Wherever the policymakers internationally were known to be considering smoke-free legislation, the experience had been that a large volume of conflicting 'evidence' seemed to emerge. Study after study on the challenging or adversarial responses of consumers and of the business community to such proposed measures, together with scientific studies on the negative economic impact of a ban or the effectiveness of softer alternative preventative options, invariably appeared when new legislation was being proposed.

A study of the surveys into the economic influence of smoke-free measures, conducted by a research team led by Michelle Scollo of the VIC Health Centre for Tobacco Control in Australia, showed that one in three such studies was directly or indirectly funded by the tobacco industry, with 94 percent of the studies claiming that such measures would have a negative impact. The researchers had reviewed ninety-seven investigations in eight English-speaking countries into the economic impact of smoke-free legislation. The funding of seven of these projects could not be identified. The team compared the quality of evidence gathered and conclusions reached, and the appropriateness of statistical tests. Their own conclusion was that policymakers could act to protect workers and patrons from the dangers of smoking in the bar and restaurant trade without any resulting adverse economic impact.

In deciding on a ban that would in effect become an international benchmark, Micheál Martin was placing a great deal of faith in advice from Irish experts – and taking into account the anticipated counterattack from members of the scientific community associated with tobacco interests. He was in effect committing the Republic of Ireland to becoming the first nation worldwide to take on such an initiative. He was also taking on Big Tobacco directly and was gambling that his resolve and his belief that this was 'the right thing to do' would succeed in the face of very well-resourced and formidable opposition from commercial interests.

So what had motivated the minister to come up with this initiative, which, as far as the public was concerned, came out of the blue, with little in the way of general debate on the issue or demand for such a measure? Why had he decided to go for an all-out ban? What about the political risks associated with the ban and the claims that this was a diversionary 'smokescreen'?

Micheál Martin explains in his own words: 'The origins of the ban go back to the Tobacco Bill of 2002, although when that piece of primary legislation was being introduced we did not envisage that a ban would go through so quickly. That bill gave sweeping powers to the Minister for Health to prohibit smoking in the workplace and indeed in any premises "the minister deems fit".

'When the legislation was being drafted, as a result of an amendment which the Opposition were looking for as well, the bill specified pubs in particular, even though they would have been covered by the catch-all phrase "premises the Minister deems fit".

'It was interesting that whereas at the time there was some limited political debate, which essentially was supportive, as was the stance of the Dáil All-Party Committee, which was in favour of a smoke-free society, the bill really did not attract any fire.

'As to it providing a smokescreen, it didn't. This was a myth. All the other health issues remained. They did not disappear. They all continued to get highlighted.

'Did I consider it risky? I actually have a passionate belief that the job of a minister is about doing things like this. Sometimes they are politically risky, and this certainly was politically and electorally risky, but it was the right thing to do. I am absolutely convinced about that.

'In November 2002, we did float the idea of partial bans and that met with negative enough comment, both from the media and from the drinks industry itself. They felt it would be impossible to police. I just floated the idea and they thought it was ridiculous.'

If that was indeed the case, it was not long before the drinks sector, most notably Irish publicans, would change their tune – in a very big way!

3

The Anti-ban Campaign Gets Going

Almost four months had elapsed between the original announcement of the ban and the first public forays of the anti-ban lobby. The opposition group which had obviously been most busy planning a strategy to challenge the ban and making campaign preparations during this period was the newly formed Irish Hospitality Industry Alliance (IHIA). Indeed, the first shots in the battle to influence the minister to change his mind were fired by the IHIA.

From June 2003, stories began to appear in the national media. Suddenly, the impact of a similar ban in New York, which had been introduced by Mayor Michael Bloomberg a couple of months earlier, in March, began to attract the attention of the Irish media. The new ban was now state-wide in New York, having previously been confined to the five boroughs of New York City. The growing interest amongst Irish media in the US ban was no doubt prompted by the anti-ban lobbyists. Stories were told of smokers in New York City seeking out Irish bars as the best places to go to sneak a cigarette with a drink. Faced with a $2,000 fine, not all Irish bars were continuing to permit smoking, but some were blatantly ignoring the ban. Also, in a parallel to the situation emerging in Ireland, dire warnings of economic chaos were being issued by the New York hospitality industry.

The other stories which began to surface highlighted instances of public disorder purportedly linked to the new habit of Big Apple bar customers taking their drinks and cigarettes out onto the sidewalk as a result of the ban. The tragic case of Dana Blake, a thirty-two-year-old security guard who died as a result of being stabbed by an angry smoker in a New York club, attracted particular attention and was used at every opportunity by the Irish bar trade as evidence of the risk to Irish bar staff. The other fear being raised was that women would have their drinks spiked if they stepped outside for a smoke.

The IHIA chose New York as the location for its first high-profile public initiative. A delegation from Ireland consisting of hoteliers, restaurateurs and bar owners boarded a transatlantic flight in the third week in July for a 'fact-finding mission'. The delegation was headed by IHIA chairman Finbar Murphy, who stressed in his first media interview that, on checking into his New York hotel, the first question he was asked was whether he would prefer a smoking or a non-smoking room. This was immediately highlighted as a potential weakness of the blanket-ban position being adopted by the Irish health minister.

The delegation had invited along a contingent of Irish journalists, as guests, in an attempt to win support for the Irish pub representatives' contention that the ban was not workable. The aim was to demonstrate at first hand the problems that existed in New York through exposing the media to the experiences of people working in the hospitality industry in the city. The visit was planned to coincide with a rally organised by the New York Night Life Association, as a protest against the Bloomberg ban. Bar owners in attendance at the rally complained that they had sustained business losses ranging from 30 to 40 percent, while trade suppliers were reported to have suffered decreases in sales of 20 percent.

Addressing the rally in Manhattan, the president of the Night Life Association, David Robin, claimed the ban was having a devastating effect and that businesses and jobs were being put at risk. His association was seeking a compromise to the ban to permit designated smoking areas in bars and greater use of extractor fans.

Encouraged by the anecdotes and the first-hand experiences they had gleaned in New York, one of the Irish delegates, Paudie O'Connell, chairperson of the Killarney Vintners Association, was reported as saying that the IHIA would be returning to Ireland with a strong message that the Irish industry would have to sit up and take this issue very seriously. 'We've learnt a lot from coming here,' Mr O'Connell stated. 'The ban is not working in New York and it will not work back in Ireland.'

The IHIA delegation was accompanied not only by journalists but also by the founder of one of the newest public relations and lobbying agencies in Ireland, Martin Mackin of the Q4 company. Mackin, a former general secretary of Fianna Fail, had, in founding Q4, joined with other high-profile media executives: Jackie Gallagher, a former personal adviser to Taoiseach Bertie Ahern, and Gerry O'Sullivan, previously head of communications at Eircom and notably linked with the controversial public flotation of Ireland's semi-state telecoms enterprise.

Questioned in New York by Conor O'Clery, the US-based correspondent of the *Irish Times,* about the financing of the IHIA delegation, Martin Mackin stated that he and Mr Murphy were financing themselves and would not accept any funding from the tobacco industry. This question about funding arose repeatedly as the campaign progressed.

As well as attending the rally, the Irish delegation was observing the outcome of a legal challenge to the smoking ban brought by the Empire State Restaurant and Tavern Association. A lobbying group called Citizens Lobbying Against Smoker Harassment also filed a suit on the grounds of discrimination against smokers.

While they were undoubtedly encouraged by what they were witnessing in the US, it was not all good news for the IHIA visitors. They also experienced a counter-rally by members of the American Cancer Society, who drew attention to new data which had been released by New York City Hall showing that, as a result of the improved environment for non-smokers, the city's bars and restaurants had enjoyed an increase in employment of 1,500 jobs. This figure was seasonally adjusted and was based on official employment statistics.

Despite this, the IHIA press-relations campaign gave great play to an announcement from the Alliance that their experts had outlined a number of scenarios that might occur as a result of the introduction of the ban in Ireland and that, under one of these scenarios, there would be 'a loss of 65,000 jobs'. Although this figure was just one of a number of hypothetical consequences that had been calculated, very predictably it was guaranteed to grab the headlines and indeed was widely featured in all coverage of this, the first serious media venture of the Irish hospitality sector. In the months ahead, the Irish media would emphasise this claim time and time again, invariably in a dismissive or critical vein. Indeed, within the hospitality trade itself the IHIA's job-loss claim was to prove to be a major tactical mistake in PR terms. Tadg O'Sullivan of the VFI was to comment: 'The figure of 65,000 job losses was a disaster which still haunts us and haunts the pro-compromise campaign in Scotland, England, Wales and Northern Ireland. It is being held up as an example of how irresponsible the pro-choice, pro-compromise campaign can be. Even when the issue of the campaign is discussed, somebody will inevitably maintain that we claimed that 65,000 jobs would be lost.'

But in the early summer of 2003, the first markers were being laid down by the anti-ban campaigners. They were arguing that the ban would have a negative impact on jobs in the Irish hotel and pub industries and that similar bans in other jurisdictions were not working.

The response from Irish media to this high-profile foray was both interesting and enlightening, bearing in mind that the event was pitched at editorial staff and broadcast producers and signalled the start of what was likely to be a long-running and highly confrontational campaign. News editors like nothing better than a good public row, so they were bound to take the story seriously.

The details of the visit and the colourful aspects of the Irish presence at the New York rally were indeed reported in all the Irish national dailies, as a hard-news story. Further feature articles appeared in a number of papers addressing the points of view of New York smokers and non-smokers alike. The 'city that never sleeps' was not found wanting when it came to providing anecdotal evidence on the impact of the ban. The comments were very mixed, however – both favourable and unfavourable toward the ban – so the *vox pop*s were essentially inconclusive. For example, one interviewee stated that 'The ban definitely affects the location of where you go out for a few drinks. I don't go downtown any more. You could not have two girls standing outside a bar on their own.'

On the other hand, a New Yorker was quoted as saying that 'It's great, nobody blows smoke in my face, and my hair and clothes don't smell like an ashtray the next morning.'

The more in-depth analysis carried out by the Irish Sunday papers was not so favourable towards the anti-ban initiative. The IHIA's claim of roughly 65,000 job losses was generally scoffed at by the media and was basically regarded as overkill. The *Sunday Tribune* political correspondent described the estimate as 'codswollop'. The *Sunday Independent* commented that, if the publicans wished to retain credibility, they should desist from outlandish claims of job losses. The *Sunday Tribune* also described the first salvo from the IHIA as nothing new to observers of Irish politics over the years. Similar scare stories about job losses had been made in relation to the reduction of Shannon stop-over flights and the ending of duty-free sales. The banning of bituminous coal was to have led to countless job losses, as was the deregulation of telecoms, not to mention the banning of smoking on aeroplanes and in cinemas. And the plastic-bag levy had originally been considered to be a complete non-starter for inveterate Irish shoppers.

The *Sunday Independent* carried a report by Jerome Reilly, who had followed up the IHIA claim about job losses by quoting from what he described as the definitive study, by Stanton A. Glantz and Annemarie Charlesworth of the Institute for Health Policy Studies at the University of California, published in the respected *Journal of the American Medical Association*. This study in effect poured cold water on all the claims of economic doom and gloom emanating from the hospitality sector in the US, in relation to the introduction of smoke-free initiatives. The main significance of the coverage of the IHIA visit was that it brought the concerns of the hospitality trade on to the news pages, and several front pages, for the first time since the announcement of the ban, back in January.

Heading into July and the start of the traditional 'silly season' for Irish media, news editors who would normally be concerned about filling newspaper pages in this 'graveyard' news period had little to worry about during the summer of 2003. The hottest topic of conversation for many years, the 'smoking ban story', was about to take off.

4

PRESSURE MOUNTS ON MICHEÁL MARTIN

Micheál Martin had a great deal of support for the ban, and it seemed to be growing. Apart from the dedicated anti-smoking organisations, other groups that rallied behind his initiative included the Irish Nurses Organisation, the Deans of Faculties of Medicine and Allied Health Professions, the College of Physicians, the College of Surgeons, the Environmental Health Officers Association – and more than two-thirds of the adult population. Indeed, by June, research polls were showing that, as awareness of the impending measure continued to grow, not only a large majority of the general public but no less than 40 percent of smokers themselves were now in favour of the smoking ban.

It was never going to be an easy task for the anti-ban campaigners, but they were certainly not short of determination and commitment, and the influence of the licensed trade in particular could not be underestimated. At the annual conference of the Vintners Federation of Ireland (VFI) in May, held in Letterkenny, County Donegal, the rallying cry to resist the ban went out loud and clear. Publicans had faced down many proposed legislative changes in the past, and conference delegates were reminded of former lobbying battles and of the politicians who had dared to challenge the might of the Irish licensed trade – and seen their careers dented as a result. The VFI were in the mood for a fight, and they fed off the memory of recent successes. After all, repeated attempts to lower the drink-driving limit to the EU norm of 50 milligrams per 100 millilitres of blood had been beaten off. They had campaigned vigorously against the Equal Status Act, which was perceived to favour the travelling community. Their major success had been in neutralising Competition Authority recommendations on licensing which would have led to unprecedented competition for existing pub-licence holders, many of whom had paid extremely large sums for licences during the Celtic Tiger boom years.

The licensed trade had to a large extent remained a closed shop, and Micheál Martin and his plans were not going to go un-challenged by a business group that had traditionally exercised a remarkable degree of influence over national and local politicians and political candidates. Every politician feared the 'multiplier impact' the local publican wielded by keeping his customers informed about local developments, and especially by making them aware of the publican's view on all local issues. Apart from cus-tomers who were in direct contact with the publican, their friends, relatives and colleagues – possibly hundreds of voters – also came within the ambit of 'the local'. The publican was considered to be the eyes and ears – and very often the banker and key supporter – of local politicians and political dynasties. In fact, publicans have often been described as Ireland's 'political landlords'.

With the campaign mantra that the smoking ban would lead to tens of thousands of jobs being lost, a massive breakdown in pub-lic order, and business closures all over Ireland, the first port of call for the anti-ban campaigners was their local political representa-tives. Initially, the main target of the publicans' efforts was the Fianna Fáil parliamentary party. Publicans began to canvass the individual members of the party to garner all possible support, with a view to building up a strong collective challenge to Micheál Martin from within his own party. A number of members of the Fianna Fáil parliamentary party began publicly to express reservations about the proposed ban and to question why Martin had not adopt-ed a more 'softly softly' approach, introducing the ban on a phased basis.

The first sign of dissent was to come from one of Martin's own cabinet colleagues and fellow Cork TD, Minister for Agriculture Joe Walsh. Walsh publicly expressed his concerns about the ban going too far. In what was to constitute the first significant Fianna Fáil brush with controversy on the issue, the agriculture minister was publicly criticised by the Irish Heart Foundation, which expressed 'extreme disappointment' with the minister for suggesting that a case could be made for designated smoking areas and for compromise. Minister Walsh was also criticised by ASH Ireland and by opposition parties.

Walsh was quickly followed into the public arena by Tipperary TD and former minister Noel Davern, a self-confessed cigarette addict. Davern at one stage described the ban as 'political correctness gone berserk' and vowed to campaign amongst his party colleagues to force Martin to back down.

Predictably, independent TDs were not found wanting when it came to jumping on the smoking ban publicity wagon, none more so than South Kerry TD and publican Jackie Healy-Rae. The colourful deputy insisted that he had received a string of complaints from his constituents about the ban, which he was convinced was 'totally unworkable'. Healy-Rae questioned whether the 'smoking police' would raid pubs and bring people up before the courts for enjoying a cigarette. He commented: 'If I went up to some fellow and told him that he'd have to go outside if he wanted a cigarette, I'd be sitting there in the bar all by myself before very long.' Later in the campaign, he was to promise to defy the ban by not enforcing it in his pub, vowing that he was prepared to go to jail if necessary.

It was clear that the hospitality industry's PR machine was getting into full stride and was working hard behind the scenes. In particular, the former Fianna Fáil advisers, now acting for the IHIA, knew that they would attract maximum attention by generating controversy and dissent amongst politicians at every opportunity.

It became impossible to avoid hearing or reading about the latest views and opinions on the ban. Phone-in programmes on local radio stations were now awash with comments on the big smoking-ban debate. On RTÉ 1's *LiveLine* radio show, an entire programme, presented in Joe Duffy's absence by Derek Davis, was made up of callers, ranging from publicans to smokers and ventilation and vending-machine suppliers, warning that the traditional Irish pub and Ireland's pub culture were doomed because of the threat posed by the ban. Jobs would be lost. Tourists would vanish. This was not the only occasion that Davis was to provide a platform for dissent against the ban. His dislike for the proposal was to continue to come through on many of his radio-programme contributions over the following months. Indeed, he was considered to have become something of an anti-ban crusader as time went on.

Not a day went by without prolonged debate on the issue on one or other of the independent radio stations, and there was no shortage of opponents of the ban who were ready to articulate the many points that had become the cornerstones of the 'anti' campaign. This was hardly surprising: a strong line-up of advisers had been brought on board to promote the message of the anti-ban campaign. The VFI was supported by the internationally linked PR consultancy Weber Shandwick FCC, which also acted for the Irish Hotels Federation, with research input from the leading market-research firm Behaviour and Attitudes. The Licensed Vintners Association employed the top-three PR agency Drury Communications, which had worked on some of Ireland's biggest PR campaigns and had brought onto their team Dublin City University lecturer Anthony Foley to conduct an economic-impact study. Q4 Public Relations not only acted on behalf of the IHIA but also represented the Irish Nightclubs Industry Association. The economic-impact study on which many of the IHIA claims on job losses and so forth had been made was produced by one of Ireland's largest management-consultancy firms, A&L Goodbody Consultants. Paul Allen & Associates PR had been hired to put the case for the Irish Cigarette Machine Operators Association, for whom an opinion poll was conducted by Lansdowne Market Research.

Collectively, the anti-ban campaigners constituted a formidable group, employing as they did some of Ireland's largest PR agencies, consultants, law firms and research companies. They were well funded and had the common aim of damaging the case for the introduction of the ban. There were signs of possible tensions between the representative bodies, however. The LVA and VFI, which were long-established and well-respected trade organisations and had always been the first port of call for the Irish media for comment on any issue relating to the pub trade, were closely monitoring the newcomer IHIA's tactics and seemed to be growing increasingly concerned about the degree of aggression exhibited in the IHIA campaign. The fear was that, if Micheál Martin found himself backed into a corner, the possibility of any compromise at some point in the future would diminish further. Comments from Tadg O'Sullivan about the activities of the IHIA confirm that all was not well amongst the anti-ban campaigners: 'From the moment the IHIA emerged, it proved to be a millstone around the necks of the pro-choice, pro-compromise campaign. They created a situation in which the minister felt that his political career depended on him forcing the total ban through, and effectively scuppered any chance of compromise. The Federation was concerned about this from the very beginning.'

At the same time, questions were being posed in the media about the sources of the funding for such an intensive campaign. Certain suspicions existed in the light of information that had emerged some years previously, in similar circumstances in the US, as a result of the leaking of a confidential memo from the Philip Morris cigarette company. This memo stressed the importance of the hospitality industry as the tobacco industry's biggest ally.

The most serious questions continued to be asked of the Irish Hospitality Industry Association, which had been specifically formed to combat the ban. This was the only group to be queried about its financial resources by the ethics watchdog, the Standards in Public Office Commission. Under ethics legislation, all contributions of €127 or more had to be declared by the recipient organisation and registered with the Commission because the campaign was now deemed to be entering into the realm of politics; no single financial contribution in excess of €6,350 could be accepted. This was only relevant to the IHIA, which was not registered with the Commission. It did not constitute an issue for organisations such as the VFI and the LVA or for the Irish Hotels Federation, whose activities were being funded by member contributions, and all of which were registered. While no spending limit for lobbying campaigns existed, the legislation required lobby groups to specify the total amount raised for any political purpose. The opposition Fine Gael Party pursued this issue, calling on the IHIA to make public the source of all its donations. The IHIA, meanwhile, registered with the Standards in Public Office Commission.

Not surprisingly, individual tobacco companies kept a low profile in the debate, although their representative organisation, the Irish Tobacco Manufacturers Advisory Committee, acting on behalf of P. J. Carroll, John Player and Gallagher, made a case against the scientific evidence promoted by the ban's proponents.

This did not mean that the tobacco industry was not extremely active behind the scenes. Dr Michael Boland, chairman of the Office of Tobacco Control in Ireland, advised an international delegation at the World Conference on Tobacco or Health, in Helsinki, which he attended in the summer of 2003, that the legal advisers to tobacco interests were already very active in Ireland and were prepared to fight the ban 'all the way' to the European Court. He informed the conference that cases were in the pipeline against the Minister for Health, the Attorney General and the Office of Tobacco Control.

What became clear at the same conference was that the response of the tobacco industry was typical of the challenges it had mounted in response to similar bans throughout the world. Another Irish speaker at the conference, Dr Fenton Howell of ASH Ireland, reminded delegates that similar doomsday scenarios were predicted every time smoking bans in various jurisdictions were implemented for cinemas, theatres and airlines, and when restrictions were imposed on duty-free sales and cigarette advertising.

Meanwhile, although collectively the anti-ban campaigners were targeting a common goal, there were growing indications of a lack of cohesion in the individual efforts of the various organisations involved. While the VFI and the LVA had common ground for dialogue and had already established positive relationships with influential commentators on behalf of Irish publicans, there were suggestions that the IHIA was interfering on their patch. Both the VFI and the LVA had a long history, whereas the IHIA was seen as a 'blow-in' operation, which appeared to be duplicating the representation of the publican sector. The tensions became evident when an economic study jointly presented to the media by the VFI and the LVA differed widely from the sensational figure of estimated job losses of 65,000 that had been announced earlier by the IHIA.

There were also a number of tactical 'own goals' which might have been avoided if the anti-ban campaign had been managed under a common 'umbrella'. Some astute media observers pointed out that it seemed to be extremely easy to anticipate and keep abreast of the anti-ban strategy, even for those who were on the other side of the campaign. At a joint presentation by the licensed trade to a large attendance of national media, Ger Nash, representing the bar-trade-workers union Mandate, circulated a printed response which was highly critical of the content of the presentation before the event had even concluded. The two sets of organisations no doubt pointed the finger at each other over who had let the infiltrator slip through the net.

Furthermore, the fact that, in order to drum up interest in the presentation, highlights were leaked to the media some days in advance may have achieved some media coverage prior to the official briefing but also had the effect of putting the pro-ban campaigners on notice and permitting them time to prepare a robust response well in advance. This particular tactical error was referred to in a number of newspapers.

On another occasion, two sharply dressed ladies, mingling in a gathering of news reporters and photographers, asked some awkward questions at a joint media briefing. The two host PR companies apparently did not realise which of the two organising teams had invited them or that the guests were from a PR company and had simply gone along to see what the event was all about, on behalf of their pro-ban clients.

These were relatively minor problems, more irritating than significant, but they served to indicate administrative and cohesiveness weaknesses in the anti-ban camp. The campaigners were not speaking with one voice and there were obvious sensitivities in relation to the key representation/spokesperson roles being fulfilled by the different organisations on the anti-ban side.

From a communications-strategy viewpoint, an interesting question arises. There were at least five separate trade-representative organisations speaking on behalf of opponents of the ban, and one of them had been established specifically for that purpose. This straight away sent signals that, while all the organisations had a common purpose, they were at the same time working to different agendas. Just as different political parties can adopt different positions when opposing a particular political issue but have to realise that their opposition is weakened because they are mounting their attacks from a number of standpoints rather than collectively, so it was with the hospitality sector.

Why did the hospitality side not achieve a greater degree of cooperation in order to speak with a more unified voice? Each of the organisations, with the exception of the new IHIA, was no doubt conscious of the expectations of its own members and would have been reluctant to surrender its particular responsibilities. Perhaps the IHIA considered itself to be an 'umbrella' group which would facilitate the collaboration of the existing organisations, but it became clear at an early stage that, for internal political reasons, the others were having none of it. Because of this stand-off, the opposition campaign was essentially fragmented and consequently was less cohesive and arguably less effective than it might have been. With a wide variety of messages and initiatives coming from the individual camps, there was no shortage of activity or comment, but there was a danger that the comments made were too diverse and sometimes contradictory. If the group had found a formula for a more collective approach, this would not have been the case.

On the other hand, the pro-health campaigners, whilst also comprising a very wide grouping of health-sector organisations and social partners, were from the outset working to a more cohesive formula. The advantage of this approach was that the 'health alliance' had a very simple objective, which was to promote the smoking ban as a positive measure to save lives. Working within the structure of a widely representative steering committee, with a centralised operational and communications strategy in place, there was a greater assurance that a consistent message would be advanced and a greater impact would be generated through collective synergy. There was also the opportunity to allocate specific responsibilities to different sectors within the group and to allow people to undertake whatever was necessary and most beneficial within their particular areas of expertise.

Notwithstanding any friction that might have existed amongst the opponents to the ban, the promotion of the campaign messages and the lobbying for support was relentless, particularly on the part of the IHIA. Their campaign, spearheaded by two former key advisers to the Fianna Fáil Party, Martin Mackin and Jackie Gallagher, was based upon exploiting every possible opportunity to generate political uncertainty or concern amongst the minister's party colleagues.

An important Fianna Fáil parliamentary party meeting was scheduled to take place in September, so with two months to go, the lobbying campaign targeting individual party members was intense. Within the party itself, there were also a number of influential anti-ban campaigners. Senator Eddie Bohan would play an important role in this respect. As a well-known publican himself, he was a former president of the Vintners Federation of Ireland and a former chairman of the Licensed Vintners Association. On the political side, he held a responsible role in the party as spokesman in the Senate on finance, with special responsibility for the Office of Public Works. As an influential spokesperson, it was natural that he would be deployed to try to influence Minister Martin. Another notable Fianna Fáil objector – and one who proved to be quite embarrassing for the government, in view of the position he held as chairman of the Western Health Board – was Councillor Val Hanley. A Galway publican, he hit out strongly against what he described as a lack of consultation on the implementation of the ban. He had put forward suggestions to the minister for non-smoking designated areas, a longer lead time for implementation of the ban, and the prohibition of smoking in areas where food was served. The fact that these comments were coming from a health-board representative and senior spokesperson made the issue particularly sensitive for the government.

Meanwhile, in terms of media coverage, the anti-ban campaigners were having their day in the sun. Media thrive on controversy, and throughout July the extent of coverage reflected the major significance the issue had assumed, with the bulk of comment, media coverage and on-the-ground activity being generated by the hospitality side.

This weight of media interest was being generated to a substantial degree on the basis that, to a large extent within sectors of the media, there was still a sense of disbelief in the possibility that the ban would come into effect. This was epitomised by the treatment of the story on radio programmes such as Sam Smyth's *Sunday Supplement* on Today FM. Smyth's guests poured scorn on the ban, discussing how impracticable it would be for bar staff in a 'tough' area to tell a customer to extinguish a cigarette. They chuckled at the prospect. Similarly, the notion of preventing an elderly farmer, coming down from the hills in the evening for a pint and a pipe of tobacco, from smoking was served up as a classic example of the fact that it would be impossible to make Irish pubs smoke-free.

Disbelief was not confined to Sam Smyth's panel, as indicated by the result of a poll conducted for the *Irish Independent* in August by TNS/MRBI. Despite widespread general support for the concept of introducing smoke-free environments, 86 percent of respondents were reported as claiming that, from a practical perspective, they believed that a total ban on smoking in pubs was unenforceable. This led to the president of the Vintners Federation of Ireland, Joe Browne, responding that this proved that the minister had not thought through the workability of the ban.

In attempting to evaluate how the campaign had progressed for both sides up to early August, the *Irish Independent* published a detailed analysis of where it perceived the parties to be positioned at that time. The newspaper credited the minister for scoring the first point in terms of achieving the 'moral high ground'. The paper warned, however, that, although he had right on his side, he faced the fate of a martyr.

Round Two was reckoned to have been won by the IHIA, on the grounds that they had recognised the influence of the licensed trade on politicians and had put forward a reasonable case for compromise, portraying themselves as an innocent injured party. In relation to enforceability, the IHIA's claims that the ban was unworkable had effectively been neutralised by the bar-workers unions' commitment to seeing the ban succeed: to date, there was no winner on this issue.

Public opinion was in favour of the ban, or certainly in favour of tackling the health issues relating to passive smoking, but it was still too early to discern the opinions of pub customers, and how these would eventually translate into commercial reality. This was especially true in relation to the idea that a smoke-free environment would encourage lapsed customers to return to pubs.

On the question of the likely negative economic impact of the ban, the *Irish Independent*'s verdict, somewhat surprisingly, was that, so far, the publicans were winning the media battle. Finally, on the issue of whether the minister would eventually yield to pressure, the paper opined that it was the nature of the Fianna Fáil Party to seek compromise and that they would probably do so again in this instance.

In light of the aggressively proactive nature of the 'anti' campaign and the sheer volume and frequency of coverage of its messages, and the perception that had grown in the media about the predicament facing the minister, it was fair to say that the hospitality trade had won the first round fair and square. The anti-ban campaigners had achieved agreement that it was imperative that, collectively, they needed to persuade Minister Martin to shift to a position of compromise. Consequently, the pro-smoking interests would now move on to develop alternative proposals to those of the minister, based on their strategic platform that they should be seen to be in favour of creating no-smoking areas and enhancing their ventilation systems.

Micheál Martin was due to take a break in August. As he packed his bags for a family holiday in west Cork, he had much to ponder. Not only had he seen the first salvo from the licensed trade as representing a formidable challenge, but he also knew that immediately on his return he would have a major obstacle to overcome in winning the support of his own party colleagues.

A great deal of pressure was mounting on the health minister to climb down from his commitment to a total ban. The big question going into the holiday period was, would Martin hold his nerve? Would the Taoiseach and his party colleagues support him and allow him to go ahead with a plan to which he had so publicly committed himself?

5

Serious Questions

from the Anti-ban Campaign

August is traditionally a quiet media month, when people with contentious messages to peddle, especially of an anti-government nature, can have a field day in terms of generating publicity. This fact was not lost on the prominent PR practitioners working on behalf of the hospitality trade.

By this stage, the newly formed IHIA was claiming to represent 3,500 publicans, hoteliers and guesthouse and restaurant owners. They had decided that the time had come to roll out a nationwide roadshow: anti-ban meetings were organised countrywide. The meetings were scheduled for Dublin, Cork, Tralee, Sligo and Galway. Meanwhile, the Vintners Federation of Ireland was orchestrating large-scale members' meetings throughout the country. Feelings in the trade were beginning to reach fever pitch.

Television coverage of these meetings highlighted the intensity of concern about the ban and the hostile reaction of the trade to it. Camera shots of tightly packed gatherings, in what were very pointedly smoke-filled meeting rooms, provided the backdrop for angry contributions from members of the trade, who in no uncertain terms demonstrated that they were not going to let Mícheál Martin have an easy ride. Much of their ire was aimed at raising doubts amongst local politicians and warning them that, if they did not support the publicans, they could expect to see mass demonstrations, refusals to implement the ban, and the withholding of taxes. There were also threats that the publican lobby would run candidates in the forthcoming local elections to highlight their cause and campaign against government interference with the 'Irish way of life'.

During the war of words, the Vintners Federation expressed outrage at the prospect of publicans facing jail terms of up to three months if they allowed customers to smoke on their premises. This was an emotive interpretation of the new law, the regulations for which were still at drafting stage; the penalties for non-compliance had not even been announced at this time. Nevertheless, the prospect of defiant publicans being sentenced to jail terms was grist to the mill for newspaper sub-editors.

Demonstrating that the hospitality sector was monitoring the situation very closely, the IHIA attempted to make an issue of an off-the-cuff general comment by Micheál Martin that there would be 'no major job losses'. The IHIA claimed that Martin had earlier categorically stated that the ban would have no impact on jobs. The IHIA was determined to undermine the minister's credibility at a time when the pro-ban campaign was highlighting the fact that 10,000 extra jobs had been generated in bars and restaurants in New York City by this stage, despite the predictions of doom.

There were also stories emerging about possible pub closures and catastrophic falls in the property values of pubs coming onto the market. The general manager of one of Limerick city's most famous pubs, the White House, promised that the pub would close its doors for the last time on 1 January if the minister went ahead with the ban. The pub had traded since 1812 and, according to the manager, had never in its history faced such a threat as the smoking ban represented. In Dublin, the Submarine Bar in Crumlin, one of the biggest 'sports pubs' and one of the largest such venues in the eastern region, was reported to be about to go under the auctioneer's hammer because the owners feared for the future of the premises, valued at €15 million, after the arrival of the smoking ban.

Next, the emphasis moved to the impact of the ban on border regions. Citing the examples which had been brought to the attention of the IHIA during their visit to New York, Finbar Murphy claimed that, just as bars on the border with New Jersey – where smoking in bars and restaurants is allowed – suffered when the ban was introduced, a similar situation would arise in Irish border towns in Donegal, Leitrim, Cavan, Louth and Monaghan. Many of these towns were famous for their huge 'ballrooms of romance' and they would suffer as customers headed north to Derry, Tyrone, Fermanagh, Armagh and Down, to continue puffing cigarettes. At a meeting arranged by the IHIA in Carrickmacross, County Monaghan, a New York bar owner flown in as a guest speaker warned the seventy local publicans in attendance that they faced the same consequences as in New York City, where, he claimed, there had been a 50 percent fall-off in business and a 20 percent drop in employment.

Every possible loophole was being exploited by the anti-ban side in order to damage the credibility and workability of the ban. The employers' representative group, IBEC, was drafted in by the campaigners to highlight the difficulty of policing all workplaces. Company cars were cited as places of work. Would the ban prohibit employees from smoking in cars provided by their employers? What about van and truck drivers? These examples were presented as demonstrating how impractical and ill-founded the new law would be. Every opportunity to denigrate the proposal was being explored and highlighted.

The IHIA drew attention to the fact that the bulk of Ireland's public houses were also the private homes of the proprietors. On these grounds, would public houses and guesthouses be exempt from the ban? The IHIA, citing figures that 7,000 of the 10,000 pubs nationwide were both businesses and homes, asserted that they had discovered a major hitch in the plan. They said further complications would arise in relation to part-time, seasonal bed-and-breakfast operations.

Next, the emphasis moved to prisons. The IHIA was among the groups to highlight the complexity of the problems that prison officers would face in enforcing the ban. The organisation appeared to be taking on the lobbying role on behalf of prison officers, who were voicing their own concerns, through the Prison Officers' Association. Nigel Mallon, the association's health and safety representative, said that his assocation fully supported the ban but that there were concerns regarding its implementation. This issue was seized on by the IHIA as another potential loophole. They were leaving no stone unturned.

Another possible hitch expounded to the media was the position of tradesmen working in other people's houses. Would workers in such a situation be exempt from the ban? The list of queries, all designed to undermine the ban, grew daily. Would households where nannies or domestic staff were employed be exempt? Would crane drivers be policed? Could pubs create smoking areas that complied with the ban, by ensuring that staff did not enter these areas? Would farmers attending marts be expected to observe the smoking prohibition? What about herbal cigarettes? Although they were not regarded as posing the same risk as tobacco smoke, how would publicans be expected to differentiate between smokers of tobacco and smokers of herbs? The fact that the minister was due to discuss specific aspects of the implementation and enforcement of the ban with staff representatives of prisons, nursing homes and psychiatric hospitals was seized upon as a sign of weakness, lack of clarity or lack of decisiveness in regard to the legal and constitutional status of the ban.

In an interview with four prominent members of the IHIA – Finbar Murphy, Killarney hotelier Frank McCarthy, Cork publican Mark Kellegher, and Paudie O'Connell from Kerry – the well-known broadcaster and *Sunday Independent* columnist Emer O'Kelly wrote that she was highly impressed by the IHIA's professionalism and thoroughness. O'Kelly cited their commitment and level of preparedness for the interview: each of the candidates had been well schooled to stay precisely on message. It was clear that the hospitality representatives had done their homework, had taken the best possible professional advice and were satisfied that they were getting across their messages in a positive manner. They appeared to be very encouraged by the way their campaign was going. They were determined to press on, convinced that public opinion was on their side. They believed that they had caused serious damage to the proposed ban through highlighting possible loopholes to a hungry media that was anxious to fill space and airtime during the traditionally lean summer period.

The hospitality sector had fired some telling shots. They had pointed out possible weaknesses in the proposed legislation. They seemed to believe they were winning the hearts and minds of political opinion-formers. On an almost-daily basis, the IHIA's advisers issued press releases and raised queries in relation to particular issues. The campaign was highly intensive, as was to be expected, since it was being orchestrated by two PR professionals, Martin Mackin and Jackie Gallagher, who in recent years had employed exactly the same media-relations tactics in helping to win support for the Fianna Fáil Party – which was now in government, and which they were challenging so intensely.

With the minister and the Department of Health team concentrating on legislative and implementation issues at this stage, the media battle was essentially being fought on the pro-health side by the non-government organisations and the trade unions. The extremely high levels of media interest and coverage being generated by the IHIA, LVA and VFI were being closely monitored and vigorously challenged by a very able team of pro-health spokespersons. Professor Luke Clancy and Dr Fenton Howell of ASH Ireland, Norma Cronin of the Irish Cancer Society and John Douglas of Mandate were in constant demand for live broadcast interviews, statements, clarifications and responses to the anti-ban campaigners. Every day, this team of spokespersons was actively pushing the pro-ban messages, not only in the national media but also on every local radio station and in every relevant print publication in the country. The hospitality sector was certainly doing the lion's share of driving the campaign in terms of proactivity at this point, but they were meeting very stern resistance all the way from a well-coordinated group of experts who were promoting positive messages in a cohesive manner.

In terms of the overall impact being achieved by both sides, the issue was one of sheer volume of coverage and proactive effort by the anti-ban side, versus very clear, consistent and fact-based campaigning by the supporters of the ban. The latter's task was to keep the focus on the very simple and emotive health message, while the protesters continually developed new arguments against the ban; in so doing, the protesters certainly generated plenty of stories and debate, but they ran the risk of causing confusion and even boredom among the general public. Indeed, research would later prove that this was the case. With the main thrust of the pro-ban campaign still at the planning stage, the role fulfilled by the health alliance at this point in the battle for hearts and minds was to prove to be of critical importance.

Apart from the hospitality side pushing their stories on the news and feature pages of newspapers, there was also an obviously carefully staged crusade of letter writing on the subject of the campaign to editors of newspapers. On a daily basis, correspondents were querying the impact of the ban on tourism and on the right of the individual, and were describing it as a cynical tactic by a government that was trying to deflect attention from the various crises it was facing. They were questioning the scientific evidence in relation to passive smoking and emphasising the doomsday scenario facing the pub trade.

It was still early August, though. Had they peaked too early and fired their big cannons too soon?

6

The Health Network Gears Up

Chris Fitzgerald, for many years the principal officer of the Health Promotion Unit (HPU) of the Department of Health and Children, had been responsible for the coordination of many nationwide health-promotion campaigns. The HPU portfolio is extensive, ranging from campaigns to tackle drugs and alcohol abuse, to healthy eating, AIDS awareness, health issues of concern specifically to women and men, and smoking cessation. The unit was behind the ground-breaking 'Nico' anti-smoking campaign, which had been applauded internationally, and it had fulfilled an extensive coordination role within the much-respected Heart Health Strategy being implemented by the Department. Fitzgerald was a veteran of many high-profile and invariably sensitive and challenging communications campaigns, many of which had been used as models for initiatives in other government departments. He was also acknowledged to be a civil servant with acute political instincts.

Against the background of the acknowledged importance of the smoking ban in terms of health benefits, as well as the highly complex nature of the project and the major political stakes associated with the ban, Fitzgerald was the logical choice to chair Minister Martin's National Smoking Cessation Steering Group, which was charged with generating the widest possible level of awareness and public acceptance of the ban. His brief was to develop a national public-information campaign that would smooth the way for the introduction of the new legislation in a manner that would be positive, persuasive and unequivocal. At the same time, the media-relations and partnership-linkage aspects of the campaign would need to be capable of responding positively to the many challenges that were anticipated from opponents. Flexibility, immediacy, consistency and, above all, credibility were to be the hallmarks of the communications programme.

Fortunately for Fitzgerald, many of the key supports for the ban were already in place. Very positive relationships already existed between the HPU and influential players in the health sector as a result of previous smoking-cessation and other health-promotion initiatives. The Office of Tobacco Control, the Irish Cancer Society, the Irish Heart Foundation, ASH Ireland, the Irish College of General Practitioners and the health-promotion departments of the various health boards nationwide were already on-side and eager to participate in a comprehensive communications networking operation.

Another timely advantage for the HPU was the fact that, as a result of an exhaustive tendering competition involving the vast majority of advertising agencies and PR companies in Ireland – and indeed other EU states – two companies had recently been selected to undertake a number of specific – and any ad hoc – assignments on behalf of the HPU. They were ready to start work on the campaign straight away. There would not be the customary delay of many weeks while a tendering process was undertaken to put the communications team in place. The advertising agency that had won the pitch was QMP Publicis, and, to the surprise of some of the more high-profile contenders, Grayling (of which the author is a former director) had emerged as the choice to handle the PR programme for the HPU. To insiders in the health-promotion sector, the selection of Grayling was not unexpected, as the agency had a formidable track record in public-information campaigns on a wide range of issues, from healthy eating to vaccination awareness and various lifestyle issues, and had been responsible for a series of award-winning 'Europe Against Cancer' campaigns. Grayling had also extended its level of expertise, having recently acquired the Curtin Communications agency, which had handled the launch of the euro in Ireland and had been working for a long period on the National Development Plan.

Rachel Sherry, the director at Grayling who had been assigned for a number of years to working on the information campaigns for the HPU, was immediately drafted onto the steering group and was given a key role in terms of coordination. With a reputation as a highly efficient and meticulous organiser – and being already very well known to the health-sector partners – Ms Sherry would find the task facing her that much easier, but there was a great deal of work to be done to ensure that all supporting organisations and potential spokespersons were 'on message'. If the pro-ban campaigners could maintain solidarity in their approach, it would immediately provide them with an advantage over their opponents, who were showing signs of fragmentation.

The core PR strategy to be implemented by Grayling was straightforward, direct and potentially powerful. It was very likely to achieve the desired result, despite the determined opposition, on condition that all the tactical elements could be made to fall into place.

Micheál Martin had put forward very convincing arguments when the smoking ban was first announced. It was imperative that, throughout the debate and inevitable controversy, these arguments would continue to remain the pillars on which the need for this action would be judged by employers, employees and the general public.

The messages about 7,000 deaths annually from cigarette smoking and passive smoking, and the importance of the ban as a means of protecting the health of workers, would at all times be central to the campaign. There would be no departure from the clear evidence put forward by the proponents of the ban. Although opponents of the ban were raising many issues, on job losses, negative effects on established business areas, and the impact of the ban on Ireland's 'pub culture', Rachel Sherry and her team set about maintaining the focus at all times on the fact that this was a very serious health issue which affected everybody, but in particular those whose health was influenced by the environment in which they worked. The smoking ban was about saving lives. The five-month period from August to the scheduled introduction of the ban at the start of January 2005 was a very long time during which to sustain interest in the core messages and to build and maintain support for the new legislation, however.

The strategy employed by Rachel Sherry, her colleague Catherine Walsh and the team at Grayling was based on enlisting the support of a very extensive network of expert spokespersons at both national and local level. The campaign would rely on creating as many opportunities as possible for the minister to repeat his main messages, but in reality this alone would not overcome the inevitable diminishing interest in the issue in media circles, where the emphasis is invariably on finding new angles. The campaign would be based on a plan to have a long list of respected health-sector experts and employee representatives available to highlight the rationale for the ban at every possible opportunity – with no deviation by anybody from that rationale. The smoking ban was a health issue, which was being advocated and supported by leading Irish health-care experts and practitioners.

Another very important element of the pro-ban publicity strategy was, at every opportunity, to broaden the debate beyond the impact of the ban on the hospitality sector. While the publican lobby had dominated the media treatment of the story to date, there was a very high level of general acceptance among the vast majority of employers and employee groups of the need for the abolition of environmental tobacco smoke. Undoubtedly, amongst workers and employers, the numbers who supported the ban very greatly exceeded those who opposed it. It was important to ensure that the media did not lose sight of this key fact.

A panel of top-level medical spokespersons was enlisted for the purpose of getting the message across through the national media. This panel would also be constantly on call to respond to or debate specific questions or to challenge comments coming from competitors. The great commitment of so many pro-ban campaigners was inspired by the unflinching position not only of the minister but also of the advisers who had originally influenced the minister to take such a strong stance on the issue.

Comprehensive briefing materials were drafted and constantly updated as each argument challenging the ban arose in the public media. Every issue was monitored, analysed and responded to. Every potential question was anticipated and deliberated. Every spokesperson was briefed accordingly.

From her previous experience in managing health-promotion programmes, Rachel Sherry was fully aware of the power and effectiveness of mobilising an extensive network of local spokespersons and linking these experts with local radio stations and the provincial press. The penetration of local media, featuring on-the-ground, local opinion, had previously proven to be critically important in forging attitudes on many national issues, and the anti-smoking campaigners knew that success or failure rested to a substantial degree on getting this media sector 'on side'. Panels of media interview candidates and spokespersons were enlisted in all eleven health-board areas. These panels consisted of GPs, health-promotion officers, nurses' representatives, union representatives, environmental-health officers and representatives of ASH Ireland, the Irish Cancer Society and the Irish Heart Foundation. These spokesperson teams, in turn, were networked with their respective local media.

A number of individuals in relevant working environments were also identified and profiled. The successful implementation of the ban on smoking in aeroplanes was highlighted by members of airline cabin crews. Bar workers were provided with platforms to welcome the clean-air initiative. Cinema owners were drafted in to dispel fears of a public backlash, based on the comparable experience of the implementation of the cinema smoking ban some years earlier. Sports personalities were on standby to support the ban. The support of entertainers who were accustomed to working in smoky atmospheres was enlisted.

Of particular importance was the identification of overseas experts who would contribute to the Irish debate by recounting the success of similar initiatives overseas. The United States in particular was considered to be an obvious source of commentators who could speak about the impact of the various bans that had been introduced over the years. For example, the Californian experience was considered to be an excellent example with which to challenge the claims that smoking bans had resulted in job losses and business failures. According to statistics from the state, the restaurant and bar trades had actually flourished in the aftermath of the ban: four years after its implementation, there were 140 more bars and taverns in the state of California than had existed there before the ban was introduced. A separate study of the impact of the New York ban, by the Cornell University Centre for Hospitality Research, had shown that significant increases in out-of-home eating and drinking and in hotel taxable sales had occurred following the introduction of smoking bans. So there was scope to draw on real experiences from the US to combat the ongoing speculation about the potential negative impact of the Irish ban.

Another major factor in the armoury of the pro-ban campaign was the opportunity to integrate the awareness campaign about the forthcoming ban with a separate campaign being run within the Health Promotion Unit to encourage people to quit smoking. Strategically, it was considered to be absolutely essential to ensure that smokers were not victimised – in other words, that it was the harm caused by the second-hand smoke rather than by smokers themselves that was the issue. It would be important to make assistance available to smokers who would use the impetus of the arrival of the smoking ban in every place of work – not just in pubs – to quit their habit. For many, the ban would be the first real incentive to quit. Advice and help would be at hand to facilitate the process, so coordination of communications was seen as being very desirable. The fact that all anti-smoking partners were working under the same 'umbrella' was of great benefit in achieving this coordination.

As with all such campaigns, timing was of the essence. The pro-ban team was acutely aware that giving almost a full year's notice of the ban could be detrimental in terms of sustaining public interest, so they had been happy to bide their time and allow opponents of the ban to make most of the early running. They believed that, as always, the last word is what matters most. The hospitality sector might have dominated the early coverage but it would at some stage soon run out of publicity ammunition and be in danger of becoming tiresome to the media. It had been decided from an early stage by the Department of Health team that the major communications effort should be focused on the last months and weeks prior to the ban's implementation. It was the classic communications strategy: the tried-and-tested way of winning support for an argument or proposition is to allow the other side to have their say, to listen to their opinions, and then to time the final response in a manner which positions it as the definitive statement on the subject at hand. Seasoned politicians, for instance, know that having the last word in a debate is of vital importance: they count down the clock in live broadcast debates in order to have the final say. The plan for the pro-ban campaign was to concentrate on the period from October to mid-December.

Prior to August, the emphasis had been on marking the initiatives of opponents of the ban and responding accordingly, but also most of the preparation for the pro-ban campaign was at an advanced stage by that time. The messages were clear; the evidence was strongly supportive of the need for the ban; the spokespersons panels were being put in place and its members were receiving media training. The next stage would be dependent on the minister winning the support of his government colleagues so that he could move the legislative process on to the next stage.

One key question which had been constantly occupying the attention of the media at this stage was the attitude of the Taoiseach, Bertie Ahern, on the issue. Whatever about the misgivings aired by certain individuals within the Fianna Fáil Party, a great deal would hinge on the position adopted by the leader. Were the former advisers to the Taoiseach, the two PR executives now working on behalf of the hospitality trade, managing to influence their former boss?

With general opinion polls showing the lowest levels of public support for the government and the Taoiseach in the life of the parliament, as a result of the perceived failure of the coalition government to deliver on its election promises, would Ahern risk further unpopularity arising from the ban? The 'anti' campaigners had painted a picture of dire job losses, tourism losses, and damage to Ireland's traditions and culture. With so many other hot potatoes to handle – including the weakening economy, ongoing revelations from the Tribunals, lengthening hospital waiting lists, problems with school funding and general unease among the party faithful, who were having a hard time at constituency level about undelivered election promises – would the Taoiseach decide that the smoking ban simply was not worth the risk?

The holiday period was still in full swing, but much was yet to come before the heavily hyped and potentially fateful two-day parliamentary party meeting scheduled for Sligo on 10 and 11 September. There were more than a few surprises still to come.

7

'SILLY SEASON' LIVES UP TO EXPECTATIONS

'I resent American political correctness being introduced to Ireland. I suppose I'm closer to Berlin than Boston on smoking, and I obviously like having a couple of cigarettes with a cup of coffee. We only have to look at New York to see the effects the ban has had. Even at this stage, Chicago is urging tourists to go there because they still have what New York used to have.' With these comments, which were widely covered in a number of media stories, the Minister for the Environment, Heritage and Local Government, Martin Cullen TD, ensured that the smoking ban captured front-page headlines in the second week of August. The silly season was getting into full swing.

As a result of some embarrassing situations arising in previous years during August, when most government ministers took a break, the Taoiseach had instructed that ministers would in future be available on a roster basis to 'mind the shop' during the holiday period. Cullen arrived back from holidays in Italy to fulfil his three-day stint on watch duty for the Cabinet and surprised everybody, while delighting duty editors, by choosing the opportunity to send off a personal broadside against the smoking ban.

As a former employee in the drinks industry – he had worked in the sales department of a drinks distribution company – Martin Cullen was aware that the smoking ban was like a red rag to a bull for the licensed sector. He was also tuned in to the rumblings of his fellow elected representatives, up to a quarter of whom were reported to be either raising objections to Micheál Martin's plan or refusing to back him publicly.

Immediately, the media began to speculate that Cullen's comments were the beginning of a heave against the Minister for Health, who was highly regarded as a serious candidate in the Fianna Fáil succession stakes. In particular, the *Irish Times* opined that this might be an attempt by Cullen, who had been achieving some worthwhile results with his environment portfolio, to put himself in the frame for a possible future leadership bid. His political career was certainly on an upward trend, but his spontaneous intervention on the smoking-ban issue caused many raised eyebrows.

Cullen, a committed forty-cigarettes-a-day man, claimed that his remarks criticising the ban 'because it smacked of political correctness' were strictly of a personal nature, but others saw the matter as a serious breakdown in party and Cabinet discipline. Despite the fact that Cullen attempted to play down the significance of his comments, claiming they were 'tongue in cheek' and 'off the cuff', Pat Leahy, a political reporter with the *Sunday Business Post,* pointed to the fact that Cullen had spoken on the issue to several national papers and had given extensive interviews on both Pat Kenny's radio show and *The Six-One News* on RTÉ. This was hardly in the manner of a glib comment! Leahy described Cullen as a cautious and canny politician who was unlikely to be making a solo run on the issue. Cullen would be expected to be particularly sensitive to the long-established links between the Fianna Fáil Party and the vintners throughout Ireland. In particular, he would be mindful of the importance of publican support for backbenchers in the local elections which were due to take place the following June.

The environment minister's intervention was exactly what the IHIA had been hoping for. Here was high-profile evidence of a possible split in opinion on the ban at the highest level in government. This dissension would now be used to mobilise support amongst backbench TDs a month in advance of the all-important parliamentary-party meeting. Heartened by this spur to their lobbying efforts, the IHIA mood at a meeting that same week in Galway was upbeat and ever more militant. There was a feeling that substantial ground had been gained. They would continue to build as much support as possible at the political level.

The IHIA bombarded political representatives with briefing materials, all pitched at heightening the concerns of politicians about the ban in relation to jobs and anti-social behaviour, as well as the impact on tourism and social traditions. Emotive messages were aimed at candidates who were singularly lacking in confidence on many fronts as they faced into the local elections. There was a common perception that publicans could individually bend the ears of the several hundred customers who frequented each of their premises. The IHIA advisers knew exactly which panic buttons to press. With the health minister out of the publicity frame on annual vacation, the IHIA knew only too well that they had a finite window of opportunity, and they were exploiting it for all it was worth.

In reality, the only significant outcome of Cullen's intervention was to ensure that the story stayed on the front pages and that it served to fuel the potential political dissension further. The environment minister himself did not emerge from the controversy in good shape. In fact, there was clear evidence that support for the smoking ban was being further galvanised as a direct result of his comments. This was witnessed at first hand at the Department of Health, with numerous calls and letters decrying Martin Cullen's stance.

Opposition parties were quick to add their criticism of his views, and the unions roundly expressed their condemnation. Liz McManus, the Labour Party spokesperson on health, said that there was now strong support for Micheál Martin to hold the line on the ban but questioned the commitment of the government to it in view of the apparent waning of solidarity amongst ministers. The Green Party claimed that Cullen's position demonstrated incoherence at government level on the issue of the ban and called on the Taoiseach and the Tánaiste to intervene and show leadership. Green Party TD Paul Gogarty said that 'the simple fact is that the health of workers must take precedence over the fears of publicans regarding a reduction in profitability.' Arthur Morgan of Sinn Féin took the opportunity to state publicly his party's position of support for the ban. Significantly, Mary Kane, of the trade union Mandate, voiced her 'amazement at the minister for the environment's opposition to the proposed ban, given the fact that his own department, in its mission statement, has responsibility for promoting sustainable development and improving the quality of life in the country.' An *Irish Examiner* editorial blasted the environment minister, stating that 'it was a crazy suggestion to facilitate the minority who smoke, without regard to the rights, or health, of the majority of people.'

It was predictable that, in the 'silly season', the media would focus their editorial comment on the progress of the smoking ban. It was time for the national press to nail their colours to the mast by declaring their own positions on the issue; this was indicative that a critical point had arrived in the context of the overall debate. What emerged very conclusively was that, despite all the protests and counter-arguments that had been proffered to date, Micheál Martin and his supporters had very effectively sold the message that this was an extremely important health issue and that the welfare of Irish people would take precedence over all other considerations – most of which were in any case driven by business considerations and profit.

The national newspapers were broadly supportive of the proposed ban. The *Sunday Business Post* came out strongly in support of it. 'There is no right to smoke in a public place, especially when doing so may put third parties at risk,' stated their editorial, which went on to describe the IHIA claim of 65,000 job losses as 'hypocritical scare-mongering' but warned Micheál Martin to implement the ban in a manner that was practical and enforceable. The *Sunday Tribune* editorial headline read: 'THERE IS NO DEBATE: SMOKING IS A DISGUSTING AND OFFENSIVE HABIT' and went on to urge the minister to stick to his guns, as victory was in sight.

The *Irish Times* carried two important lead editorials. In the first, the minister's staunch resistance against 'a determined and well-funded campaign by the tobacco, drinks and hospitality sectors' was welcomed. This was followed soon afterwards by the paper coming out very strongly in support of the ban: 'There is incontrovertible evidence that passive smoking causes serious damage to health'; 'There is evidence to show that partial bans do not work.'

The *Irish Independent* acknowledged the myriad objections to the ban that had been raised and the professionalism of the anti-ban campaign but said that it expected the 'silent majority' to mobilise and make their voices heard in the fight against Ireland's biggest killer. The utterances of the Taoiseach would be particularly relevant and were anxiously awaited. From the same media stable, the *Evening Herald,* which had been very critical of Minister Joe Walsh's call for compromise, complimented Micheál Martin for his determination to proceed with the ban.

The minister would have been heartened by the editorial stance taken by the *Irish Examiner,* which was published in his own Cork constituency but was not always the first to hand out laurels to the embattled health minister. In voicing its support for Micheál Martin on this issue, the paper stated: 'Whatever one's views of Mr Martin's stewardship in the minefield of health, he deserves applause for having the political courage not to waver in the face of an orchestrated campaign aimed at scuttling a ban based on genuine health considerations. In the coming months, as he embarks on a series of public meetings to sell his concept, Mr Martin has the comfort of knowing he is speaking largely to the converted. The political reality is that two-thirds of the electorate support his plan to outlaw smoking in the workplace.'

The tabloids were mixed in their treatment of the ban. The *Irish Mirror* denounced the zero-tolerance style of the ban, whereas the *Star* welcomed it as an incentive to people to quit smoking.

Meanwhile, the feature pages were overflowing with the topic. Paddy Murray of the *Sunday Tribune* described the ban as a cheap stunt to distract from the failure of the government to deliver on its promises. Sarah McInerney in the same paper wrote about 'nanny-state Ireland' being out of step with the rest of Europe in imposing such a ban. Her colleague, Diarmuid Doyle, attacked the IHIA and claimed that it was hard to take seriously the predictions of doom from an industry that seemed more committed to public relations than to substance, and that the anti-ban arguments lacked depth. Even the national farming paper, the *Irish Farmers Journal,* dedicated an editorial to the topic. Editor Jim O'Brien, declaring himself to be a reformed smoker, rounded on the opponents to the ban and stated that resistance to it 'should not be countenanced, let alone accommodated.' O'Brien went further: 'The drinks industry has come out with all guns blazing. PR companies and the Vintners Federation of Ireland have engaged in all kinds of spurious argument and social hand-wringing in attempts to protect their profits. There is nothing wrong with protecting profits. However, in this case protecting profits means permitting the addiction of some to endanger the health of many.'

There were so many diverging views, and so many media people anxious to declare their points of view in print, that there certainly was no silly-season news void in 2003. To sum up the situation that prevailed in mid-August: by proactively generating a huge volume of coverage, the anti-ban campaigners had won the media battle by a country mile. In terms of substance and influential opinion, however, the solid health arguments on which the pro-ban lobby were focused had put them well ahead with regard to gaining the moral high ground.

8

A Snapshot of Opinions Countrywide

Discussion about the smoking ban extended far and wide and attracted major interest amongst the all-important provincial media. Local radio stations carried endless debates, comments and reports on the ban. As far as provincial newspapers were concerned, coverage of the story was probably more widespread and diverse than on any topic since the abortion referendum.

'CLEAR THE AIR ON SMOKING BAN' ran the headline in the *Westmeath Examiner*'s editorial. The article went on to state that smoking should either be banned or not – that there should be no half-measures. The article stated that the ban should be introduced irrespective of what publicans and hoteliers felt. Another of the papers in the same region, the *Athlone Topic*, took a very direct approach, commenting that 'so serious is the damage to the physical and fiscal health of the nation which has resulted from tobacco use that there should never be any question of a government minister supporting a soft shoe shuffle approach.'

The *Kilkenny People* stated that the ban would make Micheál Martin 'fiercely unpopular' among some smokers and would end his chances of ever becoming Taoiseach, but the article commended the minister for his courage: 'Mr Martin will have long retired from politics by the time his efforts to curb smoking improve the nation's health.'

In Leinster, the People newspaper group claimed that the smoking ban would be impossible to police and called for a compromise that would be acceptable to all sides. Another paper in the same province was equally strongly opposed to the ban. The *Leinster Leader*, criticising the minister for 'being stubborn to the point of tetchiness in refusing even to discuss alternative suggestions to the sledgehammer proposal', questioned the enforceability of the ban. 'There is little point in going down the prohibition road if the law cannot be enforced. And present indications are that it may be defied in the event of a no-change attitude,' stated the paper's editorial.

Down south, the *Munster Express* questioned where the momentum for the legislation was coming from, as no such measure existed in any other state throughout the EU. The paper came down in favour of a compromise, suggesting a partial ban and a trial period. The *Longford Leader* declared its support for the ban, commending Micheál Martin for his 'bravery and commitment' and stating that it would be a poor day for democracy if his colleagues in government acceded to requests for a 'watering down' of the ban.

Despite the fact that a major share of the vocal opposition to the ban was beginning to emanate from the western region, the *Western People* gave short shrift to its opponents. The paper's editorial on the issue read: 'Isn't it amazing the noise a minority of people can make when it comes to issues like the proposal by the Minister for Health for a ban on smoking in the workplace' and went on to state that the impression was being created that the 'rights' of smokers were being portrayed as more important than those of non-smokers: 'The proposed ban is for the common good. Scientific and medical research backs it 100 percent. It's inevitable: let's get on with it.'

The *Tipperary Star* was firmly on the side of compromise: 'There is room for manoeuvre on both sides. Of course the rights of workers to operate in a clean and safe environment is of paramount importance but so too must be the fact that employment must be safeguarded.' In the same county, the *Clonmel Nationalist* expressed sympathy with the position in which the health minister found himself regarding the smoking ban but highlighted that it was even more important to tackle the twin problems of over-indulgence in both cigarettes and alcohol.

Definitive support for the ban came in the form of an editorial in the *Laois Nationalist,* which called on the minister to fight the good fight and 'Keep us healthy, stick to the smoking ban. There are lives at stake. It's quite simple: no addiction or discomfort is worth a life.' Also in support of the ban was the *Connaught Telegraph,* which stated: 'Minister Martin is right. Something has to be done to curb the death toll from smoking. It's not the end of the world and both customers and staff will be better off.' The *Meath Chronicle* threw its weight behind the ban, stating that it was a positive move and that 'In the fullness of time publicans and others in the hospitality business will surely also come to the conclusion that it was the right thing to do. No cinema closed down or airline went out of business when smoking bans were introduced.'

'AUTHORITARIANISM' bellowed the headline over the *Limerick Post* editorial column, condemning a government that was intent on banning smoking in pubs while the police said there were few crossroads where drugs could not be bought. The *Leinster Express* was not alone in coining the term 'smokescreen' on the issue of the ban, claiming that the no-smoking proposals had resulted in many other pressing issues being pushed to the sideline to some extent, as the pros and cons of the ban continued to be the burning issues in the public eye.

In Galway, the *City Tribune* expressed the view that 'If Bertie and Micheál don't go ahead with the legislation, they face the danger of being ordered to do so by the courts as bar staff from around the country begin an action to compel their employers to give them a safe workplace.' The *Western People* commented that 'Non-vested Irish society has now moved to regarding the ban as a major health issue and many smokers will regard it as an aid to breaking the habit and helping young people to avoid an addiction fostered by frequenting pubs.'

The *Midland Tribune* editorial expressed sympathy for publicans, who were faced not only with a smoking ban but also with the introduction of stringent new laws imposing severe penalties for publicans if drunkenness were to arise on or outside their premises. An earlier editorial in the same paper praised the Minister for Health on his 'commendable and courageous step. Vested interests should be faced up to and forced to back down on this issue.'

The *Galway Independent* editorialised that smoking bans were spreading like wildfire across the world and that 'For the sake of smokers, non-smokers and for our children, we must work with the minister.' In other supportive editorials, the *Meath Topic* denounced the Minister for Environment, Martin Cullen, as 'irresponsible, reckless and immature' for challenging the ban, while the *Nenagh Guardian* took the view that 'the cases put up against the proposed ban have been strong, but selfish.' The latter paper concluded that being forced to breathe in such unhealthy air stopped many people from going to the pub. 'The ban could well be the signal to bring about a lifestyle change for many, which will handsomely atone for whatever perceived revenue loss there might be,' the paper asserted. This represented one of the few positive points to be noted in media editorials in relation to the benefits for publicans. Invariably, the potential disadvantages of the ban were highlighted by its opponents and in turn were featured by the media; few commentators mentioned that the potential existed for increased business as a result of getting rid of the smoky atmosphere from hospitality venues.

Meanwhile, other potentially positive signals were emerging from the marketplace. Croom, in County Limerick, was the first town to make news headlines with the announcement that the 'Croom Mills' pub would be the first non-smoking pub in Ireland, preceding the January ban by some five months. Early reports from the new bar owners were very favourable. In the ensuing weeks, there would be announcements of other pubs following Croom Mills' example.

There was also news of the new 'café society' style which would emerge, with restaurant and bar owners planning to develop outdoor terraces and beer gardens. This was good news for suppliers of heating equipment, garden furniture and canopies. By August, orders were being taken for new purpose-built smoking shelters, while an enterprising coach company in Donegal was planning to run tours for smokers to Derry, across the border, where the ban would not apply.

Clearly, the battle for the hearts and minds of those who determine editorial policy in the highly influential provincial media had gone Micheál Martin's way. This was good news for the minister and was very important in relation to the confrontation he was about to face on his return from holidays, when he would need to focus on removing all resistance to his plan from amongst his own party colleagues.

*

National and local politicians may well have been enjoying the summer break, but at local level in particular they were also very well aware that the widespread debate at this time meant irresistible opportunities for media coverage and self-promotion. With local elections on the horizon, every opportunity for local publicity was eagerly grasped.

Local councillors the length and breadth of the land, many of whom had been earmarked for individual briefings by the hospitality PR machine, began to voice their opinions as emotively as possible. The local media were only too willing to provide a publicity platform for them during the quietest news period of the year. The councillors were also, of course, coming under increasing pressure from publicans in their communities to lobby against the ban.

Indeed, quite a number of councillors were by profession publicans: there were no prizes for guessing what their views on the ban were. For several publican councillors, there was something of a conflict of interest, because they also found themselves serving on regional health boards: they had been appointed as political representatives on the boards on the basis of their parties' electoral strengths. Ostensibly, these councillors were on the health boards to provide a voice for the general public, but on the issue of the smoking ban, surely the vested interests of these publicans compromised their objectivity on such an important health matter.

John Moloney, a Fianna Fáil TD, councillor and member of the local health board, and also a publican in Mountmellick, County Laois, said that he concurred with the minister that something needed to be done to deal with smoking-related illnesses but felt that there should have been more consultation on a possible compromise with publicans. The same views were expressed collectively by Drogheda Borough Council, which issued a letter to the minister complaining of a lack of consultation and stating its belief that there would be serious repercussions for a town that was so dependent on tourism from the North, where there were no plans for a similar ban. Local publicans had persuaded Councillors Oliver Sarsfield, Tony Ward and Tommy Hanratty to put forward a case for them seeking to have the ban introduced gradually, over a number of years. Councillor Anthony O'Donohoe said: 'I would be in favour of the ban coming in gradually, over five or six years. I don't see why there is an instant cut-off point.' Councillor Frank Maher challenged the publicans' position, however, stating: 'Smoking kills, it's not a lie, it's a fact, and I believe the minister is doing the right thing.'

Other publicans who were members of health boards and who opposed the ban included Michael Cahill, owner of the Ross Inn in Glenbeigh, County Kerry, and his fellow Southern Health Board member Tom Fleming, whose pub was in Scartaglen. Both were members of Fianna Fáil. They came in for strong criticism from their fellow health-board member Councillor Vivian O'Callaghan, owner of the Bantry Bay Hotel, who said he was appalled when six board members voted against the ban. 'I personally would like if everyone shut up and came on-side. Publicans like to get their own way, but they're not getting it this time,' Councillor O'Callaghan told the *Irish Examiner*. 'I know non-smoking areas don't work because smoke drifts. With ventilation, you put it on every fifteen minutes but people are freezing. I believe any problems with enforcing the ban will be a three-month wonder, and a major number of people will give up smoking.'

Another publican councillor who supported the ban had a more radical approach, proposing a complete ban on the sale of cigarettes. 'If you are serious about banning cigarette smoking, you deal with the supply,' said John Flanagan, proprietor of the Court Hotel in Tullamore and a member of the Midlands Health Board.

The other prominent health-board members who voiced opposition to the ban were Councillors Hugh McElvaney, a County Monaghan publican; Bernard McGuinness of McGuinness's pub in Culdaff, County Donegal; Bernard McGlinchy, owner of the Golden Grill in Letterkenny; and Askeaton publican Kevin Sheehan.

Fine Gael councillors were amongst those who refused to support their own party policy in rowing in behind the ban. Councillor Michael Fox of Tullamore described the ban as a 'diversionary tactic' and said that it was being introduced to take people's attention away from the Coalition's 'disastrous running of the economy'. His party colleague in Cavan/Monaghan, Deputy Seymour Crawford, was sceptical of the minister's awareness of the real impact of the New York ban and of the likely impact of the Irish ban in border areas: 'If Minister Martin continues on his ego trip to have this ban implemented on 1 January, I hope that he will do a tour of the pubs and hotels in the border region to explain how the law can be actually implemented in practice.' In County Cork, P. J. Sheahan of Fine Gael said he was a non-smoker 'but I advocate smoking, because tourism and country pubs will suffer without it. Prohibition is not going to work. Is the minister gone daft or mad?' he asked. Another Fine Gael councillor in Cork, John O'Shea, claimed that the ban was impinging on the civil rights of the people. 'Big brother Martin better think of another stunt, because this is not going to work', he said. The independent public representatives also had plenty to say. Cork councillor Noel Collins was reported as stating: 'My sympathies are with the elderly in rural areas who make the weekly trip to town for their pensions and stop for a drink. Then you have people who suffer from depression and see their drink and smoking as an anti-depressant.'

In Sligo, Fianna Fáil councillor and former mayor Jude Devins was quick to describe the new law as 'draconian'. He said that he believed suitable ventilation systems were available and that, in the light of his position as the manager of a local nightclub, he was of the view that owners of premises should be given the opportunity to make the necessary investment in new facilities. Athlone town councillors, including the Fianna Fáil mayor, Kevin 'Boxer' Moran, came out in support of the objections raised by Minister Cullen, claiming that the government was being over-intrusive in an issue of personal choice. The mayor said that he intended 'to bring the views of the vintners in Athlone to the Cabinet to try and add more weight to our campaign for a more relaxed form of ban.' Ironically, although a number of the Minister for Health's own party colleagues did not support him, independent Athlone councillor Kieran Molloy did, because he regarded it as a positive health move.

One councillor who cast a critical eye on the Fianna Fáil dissidents was Labour councillor Andrew McCarthy in Kerry. Condemning Michael Cahill and Tom Fleming for their opposition to the ban, he said: 'Fianna Fáil members have a great knack of criticising government policy on the one hand while continuing to support that government on the other.' He commented that his council colleagues were 'trying to be both government and opposition at the same time.'

Another party colleague of the minister's, Councillor Paddy Kiely, said: 'I can see [there being] mayhem within the pub scene, especially in rural Ireland.' In an interview in the *Carlow Nationalist,* the Fianna Fáil chairman of Muinebheag Town Commission said that, to his mind, the legislation was going too far and would criminalise people who had a life-long habit of smoking while out for a drink. 'You simply cannot change overnight an age-old culture such as people lighting up a cigarette while having a drink in a pub,' he said.

In Micheál Martin's home constituency, Cork County Council collectively voted to oppose the ban, but within hours the Lord Mayor of Cork, Colm Burke of Fine Gael, weighed in behind Martin. The minister also had the support of Fine Gael councillor Peter Kelly. In the Southern Health Board, however, opposition to the ban was particularly vocal; when it came time to vote on the issue, six members of the board voted against the ban. Predictably, considering their publicly stated opposition to the ban, these included Jackie Healy-Rae and his son Michael, who were joined by Councillors Kevin O'Keeffe, Michael Cahill, Tom Fleming (all of Fianna Fáil) and Noel Harrington. The vote was taken after a strong call for support for the ban from the board's chairman, Councillor Damien Wallace of Fianna Fáil, who said: 'We in the health board have a responsibility for the health of the people of Cork and Kerry and should back this ban 100 percent.' Board member Dr Catherine Molloy continued on the same theme, quoting George Bernard Shaw that 'liberty means responsibility', and urged fellow board members to vote in favour of the ban. This provoked a typically colourful response from Jackie Healy-Rae. Referring to the Shaw quotation, he said: 'I don't know did he smoke the clay pipe or cigarettes or did he smoke at all, and I don't care. The reality is that some people might not be alive at all but for the trip to the bar and the smoke that they had. During World War II, when tobacco was scarce, they smoked turf.'

The *Kerryman* newspaper did not hold back in roundly condemning the six councillors who had voted against the ban. 'The six men in question should all be sacked from the Southern Health Board forthwith,' the editorial blasted. 'What we have seen this week in the Southern Health Board is a disgraceful abuse of public office and the best advertisement yet for the abolition of the health boards', continued the piece, which concluded that the frenzied reaction and subsequent campaign against the proposed ban was driven by publicans and bordered on hysteria.

The divisions on the issue were rife – and across all parties the dissent was vocal, colourful and sometimes bemusing. 'Our hospitals are in a deplorable condition. Patients are being left on stretchers in the corridors because there are no beds. Worried about our health? My eye!' protested Councillor Eddie Enright, opposing the ban at a meeting of Birr Town Council. In Nenagh, the town council unanimously adopted a resolution 'calling for compromise', but only after reaching compromise on the wording of the proposal, as highlighted in the *Tipperary Star*. A doctor with the South Eastern Health Board, Councillor Tom Higgins, described the smoking ban as further evidence of a 'nanny state gone mad' and declared support for smoking areas in bars. Councillor Nora Flynn from Cappoquin said: 'We're going to jail the publicans, we were going to jail the gardaí, but we can't keep the people in jail who should be there.'

Jail was also the theme of a comment by Cork councillor Kevin O'Keeffe, who stated: 'It's ironic that you could go to jail for smoking in a pub, but you would be able to smoke legally in jail.' At a council meeting in Waterford, the point was made that councillors, who were travelling to Dublin to protest about cancer services, were now promoting smoking. In Wexford, Councillor Larry O'Brien caught the headlines by demanding a smoking facility for councillors themselves. In Carlow, Councillor Patrick Roche was reported as saying that he was glad there was going to be a ban but was not sure if he would support it, as it would constitute a further erosion of civil liberties.

Naturally, the position taken by the Western Health Board chairman, Councillor Val Hanley, in condemning the ban proved to be contentious: Niall O'Brolchain of the Green Party called for his resignation. Hanley also had his supporters, however: Councillor Michael Mullins of east Galway said that he had real sympathy for people in pubs and publicans and called for compromise to facilitate 'those people who are so desperate that they must smoke when they are socialising.' Another councillor called for something to be done about smoking in cars carrying children, instead of the focus being exclusively on pubs.

August 2003 certainly provided plenty of colourful stories on the smoking ban and filled much newspaper space. The big question was, had Minister Martin been taking heed of the many and varied views that were being highlighted in media around the country and that were becoming a preoccupation of every public representative in the country?

*

Micheál Martin may have been on holidays during August, but he was still very much on the case as far as the smoking ban was concerned. Being 'on holidays' in practice meant not making himself available to deal in person with the daily barrage of media queries on the issue.

A very able team, compromised of the minister's policy advisers, Deirdre Gillane and Christy Mannion – both of whom had played a very significant role in assessing the potential impact of the ban when it was originally proposed and had advised the minister in relation to his decision to proceed – and his battle-hardened press secretary, Caitriona Meehan, were constantly in touch with Martin at his holiday retreat in west Cork. Meehan in particular was in ongoing communication with the media on the daily agenda of issues that fall under the aegis of the Department of Health. The smoking ban was being sensationalised by every news medium in the land, but it was just another item of business for the press secretary, who was handling what was perceived to be one of the most challenging media-relations jobs of all government departments. The minister credits Meehan with holding the line when he was away from his office and in particular when, as his spokesperson, she was unequivocal in constantly commenting, in response to media queries, that there would be 'no compromise'. Indeed, Martin is quoted as saying, 'Caitriona was so forthright there was no going back after that'; he was not just being flippant.

As the mountain of issues being raised by opposition interest groups grew throughout August, the campaign communications team, coordinated by Rachel Sherry, prepared detailed responses to possible queries relating to the ban and began the rollout of expert spokespersons, who continued to focus on the health and life-saving arguments in support of the ban. A crucial point in the media's treatment of the story was arrived at on 21 August. Buoyed by their success in generating huge volumes of coverage in terms of newspaper column inches and broadcast airtime, the Vintners Federation of Ireland had announced that it would be demanding talks with the Minister for Health, with the aim of hammering out a compromise plan on his return to his daily work schedule.

'The minister cannot compromise on the health and safety of workers,' was the immediate response from the Department of Health and Children. In what was regarded as a very unusual inter-vention by a public servant, the chief medical officer of the Department, Dr Jim Kiely, moved to dismiss any suggestion that a compromise was on the table for discussion. 'What compromise can there be in this area of public health?' he asked in an interview with the *Irish Times*. In a statement, the chief medical officer re-affirmed that 'the recommendation made to the minister is that exposure to the hazards of tobacco can best be controlled by ban-ning smoking in places of work. The prohibition on smoking in the workplace is a proportionate and balanced measure to deal with the substantial risks that arise from environmental tobacco smoke.'

The anti-ban movement had made a series of high-profile objections and had generated much debate, discussion and dissent on the smoking ban issue in the preceding weeks. The response from the Department's senior medical adviser clearly indicated that these views had been noted and considered – but had not caused the minister to budge one inch. The comments from Dr Kiely were not of a personal nature: they represented the official views of the Department and the minister. 'There are no grounds for the recent conjecture that there will be any negative economic impact to this ban. Such speculation is not supported by any evidence,' stated Dr Kiely. He went on to say: 'This is a health issue, and the licensed trade should be considering the 70 percent of customers that do not smoke who may be attracted into their premises, due to the smoke-free atmosphere.'

There was no doubt about the minister's position. He was sticking to his guns and letting the hospitality sector know that their efforts to date had made no impact on his commitment or resolve. As the end of the silly season approached, much had been said and written, but the minister for health had had the last word: there would be no compromise.

9

The Fianna Fáil 'Think-in' Approaches

The Irish Hospitality Industry Alliance and publican lobby groups had failed to change the minister's mind on compromising on the ban, but had they won enough support from his Fianna Fáil colleagues to scupper his plans when they were debated at the party's annual two-day 'think-in' in Sligo in early September?

The minister and his team had not been sitting by idly. Support for his measures began to emerge in public, firstly from former minister and fellow Cork TD Ned O'Keeffe, who declared his hope that the ban would be implemented. Next to voice his support for the ban was Minister for Arts, Sports and Tourism John O'Donoghue, whose lack of comment on the issue up to this point had been notable. O'Donoghue rounded on the hospitality industry and dismissed the claims that the smoking ban would be bad for tourism. 'From soundings I have taken and research made available to me,' he said, 'I do not believe there will be any negative impact on visitor numbers as a result of the smoking ban.' Referring to a study by Tourism Ireland, the minister asserted that 'In fact, some of our people identified positive benefits in some segments': a number of US tour operators were reported to have said that there was demand in the United States for non-smoking holidays. Interviewed by Alison O'Connor and Shane Hickey of the *Irish Independent,* Minister O'Donoghue was quoted as saying that he had made his views on the proposed ban known to the Cabinet. 'It did strike me that, if John Charles McQuaid and Éamon de Valera were to return on 1 January, they would be absolutely stunned to realise that it could be easier to buy a packet of condoms than a packet of cigarettes in a pub,' he is reported to have said.

Further support for the ban was highlighted on RTÉ's *Morning Ireland* programme. Minister for Education and Science Noel Dempsey commented: 'I see this as a debate about a life-and-death issue. A number of things have already been done by the government. The ban on smoking in the workplace is a further move in that direction and I fully support it.' On the same programme shortly afterwards, Minister for Defence Michael Smith was asked whether he could see any basis for compromise on the issue. 'I don't see any great room for manoeuvre here,' he responded. 'Micheál Martin is quite right and I fully support him in looking at the causes and trying to something about these problems.'

The media were now shifting into top gear in trawling the individual deputies and senators who would be in attendance at the forthcoming party meeting in September. Reports in various newspapers indicated different levels of support and opposition to the ban, but the benchmark against which the reports were being measured was the figure of fifty dissenters, which was being widely heralded by Deputy Noel Davern, who was actively campaigning on behalf of the hospitality interests. The publicans were also aggressively lobbying for support from their local TDs. On 17 August, the *Sunday Times* carried a lengthy account of the lobbying campaign. It highlighted a meeting of forty publicans in the Brandon Hotel in Tralee, to which Kerry North deputy Tom McEllistrim and local councillor Ted Fitzgerald were 'summoned'. The report read: 'There was no shouting, no emotional scenes, no accusations. The tension was palpable nonetheless, and McEllistrim got the message. When the new deputy took the floor, he did what many of his colleagues around the country are doing – he pledged to fight an all-out ban, seeking a compromise instead.'

'Forty publicans in a small town, where pubs are an integral part of the social fabric, have a lot of power,' said John O'Sullivan, chairman of the local VFI branch and owner of the Munster Bar in Ballymullen, Tralee. 'Multiply that forty by two or three hundred and you get an idea of the number of votes at stake.'

The *Sunday Times* article went on to state that there was evidence that a growing number of deputies and senators had been swung by the arguments of the hospitality trade – or at least were giving comfort to the hospitality interests by listening to their case and, as politicians do, were appearing to be impressed by the arguments put forward. In other words, they were making the right noises to the publicans in their constituencies, but was it just lip service?

On the same day, the *Sunday Tribune* carried the results of its own poll of Fianna Fáil parliamentary members. While it had not been possible to contact all of the 111 members, the outcome was clearly at odds with Davern's claim that he had the support of fifty members. *Sunday Tribune* reporters had found widespread support for Martin's ban amongst the deputies and senators they had managed to contact. The 70 percent of respondents in support of the ban had stated that they wanted the measure implemented in full, with no half-measures. The respondents who disagreed with the ban said they did so only on the basis that they would prefer some form of compromise but nonetheless felt that the ban would be introduced, irrespective of their concerns. The newspaper held the view that support for the ban had been reinforced by an internal backlash within the party to what was considered to be the completely unacceptable intervention of Minister for the Environment Martin Cullen on the issue.

The *Irish Times* reported on the response that a combination of trades unions – Mandate, SIPTU and Impact – had received to a survey they had conducted to ascertain the views of parliamentarians. At this stage, only fifteen Fianna Fáil members had stated their unequivocal support for the ban, along with Liz O'Donnell and Fiona O'Malley of Fianna Fáil's coalition partners, the Progressive Democrats. All Opposition health spokespersons declared their support for the ban, as did Fine Gael leader Michael Noonan, Green Party leader Trevor Sargent and a number of independents.

The South Tipperary TD Noel Davern was undaunted, however. At the largest ever gathering of publicans in his constituency – 320 publicans assembled in the Anner Hotel in Thurles – Davern, who was joined at the meeting by Fine Gael deputies Tom Hayes and Michael Lowry, and Senator Kathleen O'Meara of Labour, stuck to his claim that fifty members of the parliamentary party were supporting him. The publicans left their public representatives in no doubt about where they stood, despite all the health arguments in support of the ban. 'This legislation is ludicrous and absolute madness,' stated Fred Daly, chairman of the Tipperary branch of the VFI, who said that, although he was not pro-smoking, he considered smoking to be part of Irish culture and the Irish way of life and that 'People who wanted to smoke should be facilitated as much as those who did not.' Buoyed by the combative atmosphere at the meeting and the determined speeches from the publicans, Deputy Davern left the meeting more determined than ever to proceed with proposing a motion opposing the smoking ban at the forthcoming parliamentary-party meeting.

Meanwhile, the media trawl of the views of parliamentary-party members continued, but it was notable that a number of key ministers, including Minister for Foreign Affairs Brian Cowen, Minister for Communications, Marine and Natural Resources Dermot Ahern, Minister for Finance Charlie McCreevy and Minister for Transport Seamus Brennan, did not make their views or intentions on the issue public at this stage. Media speculation continued to centre on the possible damage which might be caused to the reputations of potential candidates for the Fianna Fáil leadership who demonstrated support for the controversial smoking ban, should Bertie Ahern's ratings in the popularity polls continue to plummet as a result of a litany of problematic issues. The Taoiseach was grappling with a number of controversies in the public arena, ranging from backbencher criticism of the performance of the government, the controversy surrounding the Laffoy Commission on Child Abuse, dissent about the privatisation of Aer Rianta, concerns about the efficiency and cost of the proposed change to electronic voting, ongoing revelations from the Tribunals, and the highly publicised and widely criticised decision of his daughter to celebrate her wedding in France. He was also concerned with the likely adverse impact of the forthcoming Budget, which was being drafted against the backdrop of the bleakest economic outlook for many years. The Taoiseach was reportedly very conscious about the possible impact of the smoking ban in terms of electoral support but did not wish to elevate the ban into a major issue.

Indeed, this was also the view of various potential candidates for the leadership of Fianna Fáil: there were far more important things for the government to be doing than worrying about the consequences of making life difficult for publicans. The question still remained, however, as to whether or not the Taoiseach would support the ban – or more accurately, refrain from objecting to the way in which it was to be implemented. The reality was that the legislation already existed to facilitate the implementation of such a ban. The Public Health (Tobacco) Act 2002 was already in place, and provision for implementing a ban on smoking in the workplace could be introduced on foot of new regulations without support from the Dáil being sought. It was simply a matter of the minister signing the new regulations into law. What was at issue was the specific details of the regulations; the embattled Fianna Fáil Party certainly wanted to avoid being further weakened as a result of a public row on what they perceived to be a minor part of the government's legislative programme. For Micheál Martin, however, the smoking ban certainly was not a minor issue but a make-or-break development, because he had invested substantial political capital in it. He was well aware that he needed the public support of his leader in order to dampen opposition to the ban from publicans, the drinks industry and the wider public. From his perspective, it was frustrating that, by the end of August, six months from the announcement of plans for the ban, Taoiseach Bertie Ahern was still remarkably quiet on the subject. Had Ahern been successfully lobbied by his former right-hand man, the former general secretary of Fianna Fáil, Martin Mackin, who was now working on behalf of the IHIA?

It was the last day of August. Ministers, including Micheál Martin, were heading back to their desks after the long summer break. The Taoiseach finally decided to make his position clear. Or did he?

Speaking to the *Irish Times,* Bertie Ahern said: 'There is no compromise on the issue that smoking in the workplace is something that the government is going to ban.' He also pointed out, however, that details of the directive still had to be clarified: 'What the government have to do is finalise its work on the directive.' The Taoiseach acknowledged that many groups were lobbying the government and said that Minister Martin 'has to look at all of those points when he is making his directive and we have to try to ensure that we can find a directive that is fair and balanced.'

The Taoiseach was reported as going on to say, in the context of the hospitality industry's claims about job losses, that the ban would not result in major job losses: 'It's not just an economic argument, it's a health argument . . . and anyway I don't really believe the economic argument that everyone is going to give up drinking and eating and going into pubs.' The comments by the Taoiseach were described by the *Irish Independent* as presenting backbenchers with 'wriggle room'. Noel Davern, the most vocal opponent of the ban, sensing that the Taoiseach's position was still ambiguous, stated that he would continue to lobby for opposition to the ban among the Fianna Fáil parliamentary party and would table a motion calling for special smoking areas in pubs.

Micheál Martin moved immediately to dampen any uncertainties which might have resulted from Ahern's comments by stating that he had 'absolute confidence' that the Taoiseach would fully support the ban. Referring to the Taoiseach's comment on the fact that some details needed to be finalised, the minister for health said that he did not see any 'big issue' in relation to the circumstances of the ban. He referred to the ban on smoking in prisons and psychiatric hospitals as the only two outstanding areas of uncertainty (because both of these settings might be regarded as residences) but remained committed to introducing the ban on 1 January 'or a day or two afterwards'.

The minister was back behind his desk and, refreshed from his family holiday, was determined to move the implementation of the ban on to the next phase. Interviewed about the hostile reaction to the ban that had been so widely reported in national and local media and so actively lobbied to his party colleagues throughout August, his reaction was unequivocal. He said the claims of the hospitality sector about the economic impact of the ban lacked credibility, while their arguments in favour of smoking zones in pubs were disingenuous. 'I think it's a head-in-the-sand approach to the issue,' Martin commented, 'similar to the stance that the tobacco industry itself took when it claimed that smoking didn't cause cancer, when they knew it did.' It seemed that the Minister for Health believed that he now had clear support for the ban from the Taoiseach and was more determined than ever to press ahead with it. Martin had had many private assurances from the Taoiseach but was aware that Bertie Ahern had close connections with representatives of the pub trade and would have had extensive private consultations with them on his own initiative on this issue. The bottom line was that, although the Taoiseach had never demurred from supporting his health minister, he aimed to give the publicans a fair hearing.

What about the intense lobbying campaign against the ban, as proposed, which had been going on for weeks, and much of which was believed to have been aimed at winning the Taoiseach's support for, at least, modifications being made to the ban? How had Ahern's comments been interpreted by the vintners and hoteliers? One reaction came from John Moriarty, the chairman of the Mullingar and District Vintners' Association and owner of Mary Lynch's pub in Coralstown, who said: 'The comments are hard to follow. You would want to be a betting man to know what he [the Taoiseach] is going to say or do.' Olivia Mitchell, Fine Gael's health spokesperson, was more disparaging. She commented that the Taoiseach's response was an example of 'what he does best: come down on both sides of the argument.'

Nonetheless, the Taoiseach had finally spoken. His comments seemed to suggest that there was still room for both sides to continue to sell their propositions.

*

The Vintners Federation of Ireland executive team was beginning to come under fire from militant publicans around the country. Pub owners had begun to feel that their campaigning efforts were being overshadowed by the IHIA. Claims were made that the VFI had been too quiet to date, and too moderate. Publicans felt that the organisation should be leading the battle against the ban and not subsumed into other interest groups within the hospitality sector.

At a VFI meeting in Sligo, delegates agreed to hold off on certain actions by what the executive had termed its 'war cabinet' until after the Fianna Fáil parliamentary party meeting, but it was clear that they expected their full-time administrative team to increase its efforts.

Responding to the reported dissatisfaction of VFI members, VFI chief executive Tadg O'Sullivan stated that the battle against the ban would not be won through media coverage but at the level of the all-important government parliamentary party. This served to placate VFI members temporarily and they decided to hold off on any militant action until the outcome of the Fianna Fáil Party's deliberations became known. This was obviously a problematic time for O'Sullivan, because, as well as his members who wished to take a more aggressive stance, there were also those who simply did not take the threat of the ban seriously. He says: 'Our greatest failure was our inability to mobilise our own members or even to convince them that the ban was actually going to happen. Right up to the late summer of 2003, most of our members believed that we were the ones who were scaremongering, that the ban would never take effect, that it would never be implemented and that the public would ignore it.'

Meanwhile, the IHIA had continued to gain further valuable media coverage for its arguments against the ban. A report by consultants A&L Goodbody, commissioned by the IHIA, was released in late August. The report claimed that the smoking ban would result in lost Exchequer revenues of €1 billion per year due to people staying away from pubs. Putting forward further arguments for the adoption of a compromise to the smoking ban, the report stated that, rather than the estimate from bar-workers union Mandate of 150 deaths among bar workers from smoking-related diseases every year, the figure was in fact closer to thirteen deaths per year. It went on to claim that the average life expectancy of a bar worker had risen to seventy-seven years, in line with the national average. Another factor in the report was the estimated cost of €8.9 million which taxpayers would be obliged to pay as a result of 148 full-time staff being required to police the ban.

Mandate responded to the report by rejecting the claim on life expectancy and stating that economic evidence from other areas, such as New York, where the union had good relations with bar workers, that had introduced similar laws showed conclusively that smoke-free workplaces 'have at worst a neutral economic impact and in some cases actually helped business.'

Labour Party councillor Kathleen O'Meara dismissed the Goodbody report as 'scaremongering', stating that Labour would press for a blanket ban. Professor Luke Clancy of ASH Ireland also challenged the report, saying that 'It is easy to make claims when something has not happened. When cinemas banned smoking, there were predictions it would be the end for them, but now more people than ever are going.'

Another report soon followed from the anti-ban campaigners, this time from the Irish Cigarette Machine Operators Association (ICMOA). Based on a survey conducted by Landsowne Market Research, the report stated that 36 percent of the public regarded the ban as excessive and that 44 percent were worried that it would result in violence outside pubs. In addition, the report claimed that 70 percent of women would be anxious about leaving their drinks unattended in a pub if they stepped outside for a cigarette. At the media briefing at which the results of the survey were announced, Gerry Lawler, the spokesperson for the ICMOA, was asked if the same concerns would not apply for women visiting the bathroom. He stated that the same problem did not exist, as they would visit the bathroom less often than go outside for a smoke. The report was roundly denounced in a hard-hitting news-analysis article in the *Sunday Tribune* which stated that 'the survey by the Irish Cigarette Machine Operators Association served to illustrate the emotive, unscientific and inconclusive nature of the arguments being ranged against Micheál Martin's proposed smoking ban.'

Meanwhile, the Vintners Federation was pursuing the line of trying to influence public representatives and legislators and had made its case criticising the ban formally, in writing. The chief executive of the VFI, Tadg O'Sullivan, had challenged the fact that the advice from the minister's expert group represented a clear recommendation for a total ban. According to O'Sullivan, no such recommendation had been made, as the experts believed that extra research was needed in relation to high-risk groups such as pregnant workers and bar workers. O'Sullivan was reported as saying: 'We accept that smoking is unhealthy. We support a campaign to encourage people to quit. We cannot support scientific betrayal and false information in pursuit of that goal.'

From this point on, the tone adopted by the chief executive of the VFI was to become much more aggressive. Obviously smarting somewhat from the comments of publicans that the IHIA was taking the upper hand in the campaign, the VFI had decided to flex its muscles and re-establish itself as the leading representative organisation for publicans outside the Dublin area.

No such trade politics existed in regard to the relationship between the VFI and the Licensed Vintners Association. The demarcation lines were clearly defined. The LVA represented Dublin, with the VFI looking after the rest of the country. It was therefore perfectly reasonable that the two organisations should join forces; one of the initiatives they undertook together was to commission an economic-impact study by Anthony Foley, an economist at Dublin City University.

The findings of the new study differed widely from the figures relating to job losses and revenue impact that had been promoted previously by the IHIA. The joint VFI/LVA study claimed that between 3,100 and 8,500 jobs would be lost – the IHIA figure had been 65,000 – and that the loss to the Exchequer from diminished tax takings would be €190 million. (The worst-case scenario presented by the IHIA study had been a loss of revenue of €1 billion.)

Both the VFI and the LVA defended their projections vigorously when they were questioned about the disparity between their figures and those of the IHIA study. The VFI and LVA also disclosed survey results that indicated that 54 percent of pub customers favoured separate smoking areas rather than a total ban, with 31 percent of them favouring the latter.

Based on the publication of their report, the two publican organisations put forward a new set of proposals for a compromise to the minister's intended smoking ban. The new proposals, headed 'Customer Choice and Common Sense', were described by their promoters as realistic and operable, and as offering protection from passive smoking to both staff and customers. Essentially, the proposals were as follows:

§ Half the pub premises would be allocated for non-smoking.
§ No smoking would be allowed at the pub counter.
§ Ventilation would be improved.
§ Small pubs would be exempt from the ban completely.
§ The effectiveness of these measures would be reviewed after two years.

As a media-relations exercise, the announcement of the new proposals had been seriously jeopardised as a result of the essential components of the proposals having been 'leaked' to the *Irish Independent* four days earlier. The article, by Martha Kearns, contained not only the essential elements of the proposal but also comments from the president of the VFI, John Browne, who had agreed to be interviewed in advance of the announcement of the proposal.

There are two risks inevitably associated with leaking information to a single newspaper. Firstly, this approach may antagonise other newspapers. Judging by the fact that the *Irish Times* published a condemnation of the proposals on its front page on the day the details of the proposals were printed on page 7 of the same edition, it is clear that the tactic had misfired: the *Times* might not have run the story on the front page if it had been told about the proposal at the same time as the *Independent*. Similarly, the coverage of the proposals in the *Irish Examiner* was headlined 'PUB SMOKING PLANS REJECTED', with the *Star*'s banner declaring 'MARTIN DISMISSES BAN REPORT'. Secondly, the leak had given the minister's team the opportunity to prepare its response to the proposals well in advance of their announcement. Minister Martin issued a strong rejection of the proposals hours before they were formally announced, and the pro-ban campaign's media-relations coordinator, Rachel Sherry, had prepared for equally emphatic – and dismissive – responses to be issued from the trade unions and from health and ventilation experts.

In a number of media reports, Micheál Martin highlighted the fact that there had been adequate time – almost a full year from the date of the announcement to its proposed implementation – for employers to adapt to the new regime and that no purpose would be served by a further two-year delay. Rejecting the idea of a designated smoking area, he said that 'Smoke permeates the entire atmosphere of a room regardless of where the smoking takes place, without regard to areas which are designated smoking or non-smoking.' He asserted that this was not a realistic option and reaffirmed his belief that 'There is no safe level of exposure to this carcinogen.'

Support for the main thrust of the VFI/LVA proposals came from the IHIA, who had added a new dimension to their PR campaign by enlisting rugby legend Peter Clohessy to their publicity team. As the owner of the Sin Bin pub and nightclub in Limerick, 'The Claw' was drafted in to address publican meetings and take part in media relations. Apart from restating the threats to jobs and volumes of business, he joined in the call for separate smoking areas to be created and for smoking at pub counters to be banned. On and on rolled the anti-ban PR machine. With only days to go to the Fianna Fáil meeting and with silly-season news space still available to be filled, there was no let-up in the flood of press releases, counter-proposals and suggestions.

On 22 August, the employers group IBEC was persuaded to add its voice to the debate. IBEC called for more research to be undertaken before the ban was implemented and stated its preference for designated smoking rooms in workplaces. The *Irish Times* reported Tony Briscoe, assistant director of social policy at IBEC, as claiming that the ban would be 'hugely problematic' for businesses, with individual firms likely to face costs of €3,000 each if they complied with new regulations. The *Irish Times* described IBEC as supporting the IHIA's position and noted that hoteliers and guesthouse owners were members of the employers group.

As the PR campaign against the ban progressed, newspapers reported accusations from Dr Chris Proctor, the head of science at British American Tobacco PLC, that the health authorities in Ireland had compiled their scientific information in an unfair and biased manner. These comments, contained in a submission by the Irish Tobacco Manufacturers Advisory Committee (ITMAC) to the Health and Safety Authority, suggested that the new legislation should respect the preferences of smokers and non-smokers alike. The ITMAC submission stated: 'ITMAC believes that current regulatory proposals to extend workplace smoking restrictions to include social settings such as pubs, clubs and restaurants are not based on sound science and there is insufficient evidence either in Ireland or internationally to support a total ban.'

It would appear that the urgings of the tobacco group did not sway the Health and Safety Authority. Within days, reports appeared of a recommendation being prepared by the HSA which would go to the Tánaiste, Mary Harney – who was also, as Minister for Enterprise, Trade and Employment, responsible for the HSA – seeking a designation of second-hand smoke as a 'cancer-causing agent'. Such an official designation would be a critically important milestone in the overall process.

Frank Fahey, the minister of state with responsibility for labour affairs, was being advised that the new designation of passive smoking would lead to all employers being compelled to take every possible step to eliminate or minimise the exposure of employees to smoke-related risks. The minister was told that employers would be issued with guidelines advising them to take steps in this area in advance of implementation of the ban on 1 January.

Meanwhile, not all of the media coverage was emanating from the anti-ban side. Proponents of the ban continued to press for support for the initiative, albeit in a more measured and less sensational fashion. For example, Irish physicians involved in cardiology and cardiac surgery urged Minister Martin not to give in to vested interests. Dr Mahendra Varma, the president of the Irish Cardiac Society, stated that she viewed the harmful effects of passive smoking as an extremely serious health issue. 'It is now incumbent on the minister, and indeed the government, to protect workers, based on the medical information available,' Dr Varma said. 'Thousands of patients occupy hospital beds in this country each year [as a result of] diseases directly related to the effects of tobacco smoke.' She urged the government to introduce the ban 'without compromise or dilution'.

The Office of Tobacco Control (OTC) also moved to clarify some queries in relation to implementation of the ban in particular settings, such as psychiatric hospitals and prisons. Tom Power, chief executive of the OTC, announced that special codes of practice were being drawn up for certain institutions and that separate codes would be published for pubs, hotels and guesthouses. This was confirmed by Minister Martin, who stressed that a balance had to be found in order to protect the rights of employees, patients and inmates of these institutions.

As the politicians packed their bags to head for Sligo for the first parliamentary-party debate after the summer recess, they had much to think about. They had been exposed to an intensive lobbying campaign against the ban and widespread media coverage of all the arguments in the relentless media treatment of the story throughout the summer of 2003. They were conscious that media circles believed that a critical juncture in the smoking-ban campaign had been reached.

The annual 'think-in' of the parliamentary party took place over two days, 10 and 11 September. No motion against the ban was tabled. Indeed, Deputy Noel Davern did not even attend the meeting, reportedly opting instead to be part of an official delegation visiting Poland and Croatia that week. To judge by media reports, it is doubtful whether the ban was even discussed at the think-in, other than in the context of a workshop on health issues in general. Instead, those in attendance were told to focus on 'getting back to basics' and not to air their grievances publicly, to the detriment of the party.

If August traditionally represents the silly season, August 2003 was exemplary in this respect. And there was a great deal more to come.

10

Public Support Wanes,

Publican Resolve Increases

On 15 September, the *Irish Independent* published the results of a poll conducted by Millward Brown IMS, which showed that public support for the ban had fallen to 52 percent. Worryingly for the government, just 37 percent of respondents were in favour of the timing of the introduction of the ban, on 1 January. Brian Dowling, the paper's political correspondent, suggested that the results of the survey would give renewed momentum to opponents of the ban, who would interpret the findings as an indication that their campaign was making a positive impact.

The *Irish Independent* survey was followed soon afterwards by a similar poll, conducted on behalf of the *Irish Times* by TNS/MRBI, which indicated a modest decrease in public support for the ban, from 59 percent in February to 56 percent. The proportion of those polled who disagreed with the ban had risen from 36 percent to 40 percent. Amongst the findings that would have encouraged the ban's opponents were the figures showing that 49 percent of those in the eighteen-to-twenty-four-year age category were opposed to the ban, with 47 percent in favour. It also showed that, in the less-well-off C2DE social group, more people opposed the ban than supported it, by a margin of 50 percent to 46 percent. In the ABC1 group, however, a very healthy 64 percent of respondents were still in favour of the ban.

Rather ironically, the suggestion that public support for the ban was to some extent waning served to reinforce the Taoiseach's resolve on the subject. The fact that there was still a very strong body of public opinion in favour of the ban, despite the sustained onslaught over the preceding two-month period from its opponents, seemed to galvanise senior politicians. Questioned by the *Irish Independent* about whether declining support for the smoking ban would cause the government to change its stance, Bertie Ahern insisted that the ban would go ahead in January, as planned. A clear message was being sent to TDs and senators that the government was not for turning.

The Taoiseach's position hardened further a few days later, when he embarked on a five-day official visit to the US. While he was in New York, Ahern visited a number of bars and spoke to people in the catering industry. Various media reported him as commenting that 'Most people, as far as I can see, appreciate the clean, healthy air and feel the ban is achieving its purpose. It has settled down well.' The Taoiseach was told by New York City officials that opposition to the ban, from Irish-American bar owners and others, was beginning to run out of steam.

Health minister Micheál Martin also visited New York at the same time as the Taoiseach. Martin was in the city to sign the World Health Organisation's international treaty on tobacco control, at the United Nations headquarters. He also used his visit to see for himself the impact of the smoking ban in New York State and to meet Mayor Bloomberg. 'Mayor Bloomberg told me there had been an increase in applications for liquor licensing since the ban,' Martin said. The minister also stated that, based on information he had received from official sources, it was his view that the ban had not had a negative economic impact; he said he had come away from the visit with his mind firmly made up to proceed with the ban on 1 January, as it was the correct step to take. 'The experience within the restaurant that is smoke-free, or indeed a bar, is superb,' he said.

Minister Martin also included a visit to Washington, where he met US Health Secretary Tommy Thompson and Senator Jack Reed of the Senate's Health Committee, and visited the National Cancer Institute. At the conclusion of his visit, the minister announced that an element of the Irish ban which he was keen to encourage was that, to some extent, it would be 'self-policing'. With this in mind, he said he would consider setting up a free-phone 'hotline' to enable the public to report breaches of the smoking ban. In an interview with Conor O'Clery of the *Irish Times* prior to his return from the US, Martin said that he was encouraged by the 97 percent rate of compliance with the ban in New York and that he was more committed to the ban than ever. He said: 'Environmental tobacco smoking is a cause of cancer. It is the same as exposing workers to asbestos. I genuinely feel I have no choice but to act to protect the health of workers.'

By a strange coincidence, at the very time that the minister was meeting Mayor Bloomberg in New York, the latter's predecessor as mayor of New York, Rudi Giuliani, was visiting Ireland. Giuliani, who had risen to international prominence as a result of the leadership he had shown in the aftermath of the September 11 terrorist attacks, was guest speaker at an MBA Association of Ireland conference held in Cork. He used the visit to offer his views on the impact of the New York ban. In his opinion, the ban 'was not fair to smokers, who should be provided spaces of their own. Some people want to make the choice of being able to have a cigar or a pipe or a cigarette after dinner, and they should be provided with an opportunity to do that.' During his term of office, he had introduced the first controls on smoking in bars and restaurants, in 1994. These controls permitted smoking in designated areas, which had to be ventilated to ensure that smoke did not affect people who were eating or drinking in non-smoking areas. 'I thought that was the best way to achieve a balance,' he said.

Meanwhile, the Licensed Vintners Association was busy publishing the findings of a survey of its own, this time canvassing the support of bar staff. The LVA claimed that the majority of workers in bars throughout the country were in favour of its compromise proposals, and that this would be proven by their survey. This prompted the trade union Mandate and the Irish Congress of Trade Unions to publish a rebuttal of the LVA's assertions, on the grounds that individual workers would feel intimidated by being asked by employers to sign a petition supporting a stance that had been taken by their employers. The unions went further, highlighting the fact that, by signing such a petition, workers would be effectively acknowledging that they were in favour of working in an environment that was potentially harmful to their health.

*

At the same time, the VFI was also carrying out a study of the effectiveness of ventilation systems. The study was being conducted on behalf of the VFI by the University of Glamorgan. Tadg O'Sullivan of the VFI claimed that two-thirds of VFI members throughout the country had invested in ventilation systems. He anticipated that the results of the study would confirm the view – which he had been promoting very forcefully – that modern ventilation systems were effective in dealing with secondary tobacco smoke. He said that publicans would fully support a change in licensing laws compelling them to install high-grade ventilation systems and that he believed that the study would be a key factor in persuading the government to climb down from its commitment to an all-out ban.

The publicans chose to step up their emphasis on the ventilation issue in response to an EU study which found that increased ventilation had little impact on reducing the effects of smoking in bars and restaurants. This study had served to support the views of the minister's expert group, which had earlier concluded that ventilation technologies, including air-conditioning systems, could not adequately control worker exposure to environmental tobacco smoke. The most advanced technologies, such as displacement ventilation, had the potential to reduce levels of environmental smoke by 90 percent – but this still left exposure to hazardous air pollutants at 1,500 to 2,000 times acceptable risk levels, the study reported.

Tadg O'Sullivan of the VFI said that the EU study was invalid because it was based on ventilation equipment which did not meet the standards of that used by his members and as set down by the Chartered Institute of Building Service Engineers. This standard provided for ten to fifteen air changes per hour, instead of one per hour, which he contended was the standard on which the EU report was based. According to O'Sullivan, properly conducted and verified tests using modern equipment would show that the EU study was without merit. His criticism of the study was not helped, however, by statements from the Philip Morris website, carried by the *Irish Examiner* in a report by Dan Buckley, which suggested that, while ventilation was capable of reducing the sight and smell of tobacco smoke, 'It is not shown to address the health effects of second-hand smoke.' Philip Morris asserted that 'Ventilation is merely the dilution of unwanted indoor air constituents (such as smoke or odours) with fresh outdoor air.' The website added: 'Philip Morris USA believes that the conclusions of public-health officials concerning environmental tobacco smoke are sufficient to warrant measures that regulate smoking in public places. We also believe that, where smoking is permitted, the government should require the posting of warning notices that communicate public health officials' conclusions that second-hand smoke causes disease in non-smokers.'

Closer to home, studies on the impact of environmental tobacco smoke (ETS) in Ireland were influencing the scientific community. In Galway, children from three schools were monitored by Maurice Mulcahy, senior environmental health officer at the Western Health Board, and Dr David Evans of the Department of Public Health. The study showed that children from homes where family members smoked had more than three times the levels of cotinine – the breakdown product of nicotine – than children in smoke-free homes. The study concluded that children were regularly exposed to the dangers of ETS in their own homes (if family members smoked), in other people's homes, in restaurants and in cars.

*

Whatever about the intense arguments regarding the harmful effects of smoking indoors, the problem of cigarette ends discarded outdoors by the people of Dublin had begun to exercise the imagination of the city's colourful new Lord Mayor, Councillor Royston Brady. Worried that the ban on smoking in pubs would lead to an escalation of the existing problem for Dublin Corporation of clearing up the butts – at the time, it was collecting

624 million cigarette butts every year, or one every twenty seconds – Brady announced a new disposal scheme. In order to highlight the fact that discarding a cigarette butt would now result in a fine of €125, the lord mayor announced that 20,000 portable ashtrays would be handed out to smokers in Dublin free of charge. These fire-proof boxes could be carried in a pocket or purse; new ones could be purchased for €2 apiece.

The lord mayor's 'Butts Out' campaign was lambasted in the media as 'flawed thinking' and 'providing an incentive to smoke'. Probably the only occasion that the boxes were seen in public in significant numbers was at the predictable launch photo-call, which featured the lord mayor surrounded by a team of female models. The episode provided a little light relief in the context of the overall communications struggle on the subject of the ban.

*

As debate on the ban grew and grew in Ireland, the interest in the proposed legislation from overseas intensified. EU Commissioner David Byrne came out with a proposal that the initiative being introduced in Ireland would be beneficial if it was rolled out across the EU. Byrne accused the tobacco industry of deliberately targeting young people and said that it was important to put health before profits. He endorsed the Irish intention to introduce a workplace ban and said that the EU Commission was considering other measures to combat tobacco smoking, including increased minimum-level Europe-wide taxes on cigarettes and measures to develop alternative crops for European tobacco growers. While the Commission did not have the authority to ban smoking in bars and restaurants, it could achieve beneficial results through worker-protection legislation and a ban on tobacco advertising and sponsorship.

In the UK, Prime Minister Tony Blair was coming under pressure from British medical interests to introduce an Irish-style ban, the debate on which was being closely observed from across the Irish Sea. Blair was seen to be reluctant to take a stand on the issue at a time when he was under intense pressure over Iraq and problems with public services. Similarly, his health secretary, John Reid, a former heavy smoker, was not inclined to move to what he termed 'draconian measures'. In the UK, the politicians were hoping that commercial interests would tackle the problem voluntarily, such as by introducing more smoke-free areas in pubs and restaurants, requesting that customers not smoke at the bar, and so on.

The Italians were also up in arms over Commissioner Byrne's proposals to coax farmers out of tobacco growing through amendments to farm subsidiaries. The Italian agriculture minister, Giovanni Alemanno, was less impressed with affairs in Ireland, which had sparked off the debate, than with putting pressure on his fellow countryman, Commission president Romano Prodi, to block such an initiative.

Jeffrey Wigand, a former employee of the US tobacco giant Brown & Williamson, was also observing the debate on the Irish initiative. Wigand had been fired and subsequently sued by Brown & Williamson for stating publicly that the company knew that nicotine was an addictive substance. He later became a spokesperson for the non-profit organisation Smoke-free Kids and advised governments on anti-smoking measures. His experiences were used as the basis of the 1999 Hollywood film *The Insider*; he was played by Russell Crowe, with Al Pacino co-starring as the persistent TV producer who persuaded Dr Wigand to take part in an exposé interview, which was controversially shelved by CBS due to fears of litigation. Wigand was reported in the *Irish Examiner* as stating that 'Ireland has shown an enormous leadership role in making every place smoke-free, including pubs.' When he was interviewed by Matt Cooper of the *Sunday Tribune* and Today FM, Wigand laughed at the claims being made by the VFI about ventilation. 'Just look at the Philip Morris website and see what the tobacco giants think about the dangers of environmental smoke,' the most famous whistle-blower on the tobacco industry commented.

*

On his return from the US, the Minister for Health was to find on his desk an eight-page letter from solicitors for the VFI. Tadg

O'Sullivan announced that the federation had been examining, from a legal perspective, the contents of the independent report on which the minister had made his original decision to introduce the ban. The VFI claimed that there were serious flaws in the report, and the organisation had written to the minister outlining what it regarded as 'inconsistencies, discrepancies and inadequacies' in the report. O'Sullivan said that the VFI would consider taking a legal challenge to the ban: 'If we find, legally, that the report is flawed and the minister takes action based on that report, then we're entitled to consider anything.'

Would Minister Martin hold firm? The only concession on the introduction of the ban was a comment by the minister that the ban would come into force 'in the first couple of weeks of 2004' rather than on 1 January. Was this a signal that all was not well in the pro-ban camp?

It was now the end of September: people were back at work, and it was all systems go on the smoking ban campaign trail.

11

THE PUBLICANS SEEK TO 'DIVIDE AND CONQUER'

'FIANNA FÁIL RIFT OVER SMOKING BAN'; 'DISSENT DEPUTIES TO FAN FLAMES TO REVOLT'; 'FIANNA FÁIL LOBBY FOR DEAL ON SMOKING BAN GROWS'; 'FIANNA FÁIL FACING TOUGHEST ARD-FHEIS IN YEARS'. The headline-writers were unconvinced that Taoiseach Ahern and Health Minister Martin had put the smoking-ban issue to bed as they headed into the new parliamentary session. It was early October, the Dáil was back in business, and the Fianna Fáil ard-fheis was scheduled to take place in Killarney on 11 October.

Fine Gael leader Enda Kenny seized on the rumblings that were emanating from Fianna Fáil objectors to the ban by raising the matter in the first session of the new Dáil. Twenty members of the Fianna Fáil parliamentary party had reportedly sought a compromise solution at their first meeting since the Dáil resumption. 'It appears that the smoke signals coming from the Fianna Fáil workplace seriously undermine the government's decision,' Deputy Kenny commented. He went on to press the Taoiseach to clarify whether there would be any compromise in relation to the ban; Ahern replied, on the Dáil record, that 'I will reiterate what I said: from January, smoking will be prohibited in the workplace.'

Dissent on the smoking ban within government circles was still definitely an issue, however. The campaign by the publican lobby may not have delivered the desired result at the party 'think-in', but the seeds of doubt and uncertainty had nonetheless been sown. That was enough for the anti-ban groups to continue with an intensive lobbying campaign. After all, despite the rallying call to focus on the more significant economic and social issues and to highlight the past successes of the Fianna Fáil/PD coalition government, there was an unprecedented backlash from the public to deal with, as evidenced by the opinion polls. The publican lobbyists identified this public disenchantment with the ban as their most valuable weapon in motivating more Fianna Fáil Party members to support their cause.

The lobbying campaign was based on feeding the concern that the smoking ban was an unnecessarily harsh piece of legislation which should be modified along the lines of the compromises that had been put forward by the VFI and others. The anti-ban campaign aimed to persuade deputies and senators that with the public finances being inadequate to meet the promises made in advance of the last election, a variety of measures being introduced which were seen as inhibiting personal freedom, and even the deteriorating situation in Northern Ireland jeopardising the Good Friday Agreement, the smoking ban was a very unwelcome additional problem for public representatives to have to deal with.

Sean MacCarthaigh, political correspondent of the *Sunday Business Post,* summed up the mood of some unnamed Fianna Fáil members: 'It's like, you can't smoke in the pub or you'll be arrested. You can't drive fast even when it's safe or you'll be put off the road. You can't park your car or you'll be clamped. And if [Minister for Justice Michael] McDowell has his way, you can't have a late drink.' Another commented: 'Once you get the reputation as killjoys making life difficult for ordinary people, it's very hard to shake it off.'

The mood within the Fianna Fáil Party was decidedly downbeat: it was heading into the last session before the added distraction for the government of the six-month presidency of the EU, which would preoccupy all senior members of the party. The anti-ban lobbyists hoped to further their cause within this very receptive environment. With the all-important ard-fheis scheduled for early October, their mission was clearly to work on developing the divisions in the party as much as possible. This strategy of 'divide and conquer' would be the last major throw of the dice for the hospitality sector. They determined to step up their campaign to its highest point since the announcement of plans for a smoking ban and were encouraged by some high-profile support within Fianna Fáil.

The *Irish Times* was the first newspaper to name names: political correspondent Mark Hennessy reported that two ministers of state, Pat 'The Cope' Gallagher and Frank Fahey, were amongst those seeking compromise on the ban. Both ministers were reported to have spoken, at a parliamentary-party meeting, in support of a motion seeking changes to the proposed ban. It was remarkable that Fahey should have been one of the many speakers who were reportedly in favour of compromise, in view of his position as minister with responsibility for health and safety in the workplace. As such, he would have had first-hand knowledge of the significance of the ban and of the reports and advice which had prompted the Minister for Health to adopt his unequivocal position on the issue. As Fahey would also have direct responsibility for enforcing the ban, his apparent lack of support for its introduction in its original form was regarded by many as unfathomable. In fact, he was under pressure from Galway publicans on the issue. Further fuel was added to the controversy by comments from a leading publican in the Galway area, Christy Ruane of the VFI, who claimed that Minister Fahey had made a commitment to local publicans to support the call for compromise. Sylvester Cronin, the safety officer of the trade union SIPTU, responded to the reports of Minister Fahey's opposition to the ban by saying that 'It seems to me that the minister is acting contrary to his ministerial obligations if he is prepared to compromise on the health of some workers instead of protecting the health of all workers in the State.'

The Fahey issue was to occupy, albeit briefly, the attention of both the national media and the local press. This episode reached almost farcical proportions shortly after a statement was issued on behalf of Fahey, proclaiming his support for the health minister and stating that he had 'always supported government policy'. This drew a further response from Christy Ruane, who said that Fahey 'is beginning to double-deal, and that is not a nice way to conduct politics. He even voted against the ban at last week's parliamentary meeting.' There were also claims that Fahey had met cigarette-industry interests and had indicated to them that there was room for compromise.

The media set about tracking down Minister Fahey to seek

clarification of his position. It emerged that Fahey had travelled to Spain on a golfing holiday immediately after the parliamentary-party meeting, little knowing that he had become the centre of attention in the biggest media story of the year. On his hurried return, at the request of the Taoiseach, to Dublin Airport, he found himself greeted by an army of reporters, television cameras and photographers. The clearly unsettled minister was at pains to reiterate his support for the ban but voiced his concerns about certain enforcement issues and stated that he was awaiting recommendations from the Health and Safety Authority. 'We need to have extra resources in order to implement this ban effectively,' he claimed.

Minister Fahey's absence on a golfing trip at such a critical stage in the debate about a piece of legislation with which he was directly involved did not escape the typically ascerbic eye of satirical-feature writer Miriam Lord of the *Irish Independent*. She wrote: 'Yesterday afternoon, as if by magic, a written statement arrived in from Frank. Glory to God, but not even Tiger Woods on fire can pitch up a delivery from Spain to Dublin. Not that the statement gave any clue to the junior minister's whereabouts. He just wanted to make clear that he was full-square behind the smoking ban and that Bertie was going to break his legs if he didn't get home immediately and say that in person.'

The message from this episode to politicians is that, if you are involved in controversy, stay put, cancel the trip to the golf course and, most of all, don't expect that issuing a press release before you depart on holidays will get you off the hook. It is better to stand your ground and express your own views clearly and in person.

*

The hospitality sector's campaign of publican meetings country-wide was resulting in growing numbers of public representatives stating their support for some form of compromise, and the Frank Fahey episode did not deter a number of other ministers of state from going public about their concerns regarding the smoking ban. There was a great deal of media interest in the public declarations by three ministers of state – John Browne, Liam Aylward and Michael Ahern – that they believed that Minister Martin should soften his hard line on the ban. Their comments were reportedly prompted by a feeling amongst many of the attendees at the recent Fianna Fáil parliamentary meeting that, if a motion proposing a compromise had been put to the meeting, it would probably have been adopted.

Defending their positions in interviews published in *Ireland on Sunday,* the three junior ministers all said the same thing: that adjustments to the proposed law and a phased introduction were necessary. John Browne, minister of state at the Department of Communications, stated that he had consulted with all publicans in his Gorey 'patch' and had put it to them that designated smoking areas and proper ventilation were preferable to an all-out ban. Michael Ahern, minister of state at the Department of Enterprise, Trade and Employment, stated that, although he favoured the restriction of smoking in public places, the new measure should be phased in gradually. Liam Aylward, minister of state at the Department of Agriculture, said that he believed that in its proposed format, the ban would be unenforceable: 'Small rural pubs just don't have the wherewithal to deal with this. I'm in favour of the ban in principle but I just want people to be practical about how it is introduced.'

Backbenchers also ensured that their views were made public. One of the most outspoken was County Limerick TD John Cregan, who claimed that the ban was 'too much, too soon'. Cregan, a smoker himself, commented: 'Rightly or wrongly, this ban is perceived by the public as a police-state action, and we have to listen to what the people are saying.' Stating that the government was 'digging itself into a hole', Cregan claimed that similar views were held by 'every backbencher in the parliamentary party.' Another of the backbenchers who challenged the ban was Longford Fianna Fáil TD Peter Kelly. 'My main concern about this blanket ban is that the ordinary countryman or woman who enjoys his or her pint in the local in the evening will be discriminated against,' he said. Deputy Kelly, a former publican, said that he was worried that a blanket ban would cause job losses in the hospitality industry. Another former publican, Deputy John Moloney of Laois/Offaly, publicly called for compromise. Even Deputy John Moloney, the chairman of the Fianna Fáil Health Committee and a member of the Midland Health Board, publicly aired his support for delaying the introduction of the ban.

An ominous development came from Fianna Fáil's partners in the coalition government, the Progressive Democrats, when junior minister Tom Parlon challenged Minister Martin to engage in further consultation with hoteliers and publicans on the ban. In his capacity as minister of state in charge of the Office of Public Works, Parlon attended a meeting in Kerry at which members of the Irish Hotels Federation expressed outrage at the introduction of the ban. Parlon felt obliged in a subsequent interview on Radio Kerry to call on the Minister for Health to take on board – or at least listen to – the views of the hospitality industry.

*

Just as the hospitality campaign was regarded by many to be reaching its peak, the pro-health supporters were putting the final touches to a number of key initiatives aimed at maintaining their lead in public opinion on the issue of the ban and in particular to ensure as widespread a welcome for the ban as possible in the period immediately prior to the announcement of the regulations governing the new law.

The momentum had certainly been captured by the pro-smoking campaigners in the weeks and days leading up to the Fianna Fáil congress, but there was a strong belief that many of the politicians who had voiced their support for compromise might have done so primarily to show that they recognised the concerns of their constituency publicans. At the end of the day, they would most likely toe the party line, it was felt.

To date, those same politicians had not had an opportunity to demonstrate their support for the ban openly, so, in a cleverly devised plan to counter the much-publicised attendance of politicians at hospitality-industry events countrywide, Jane Curtin, communications manager at the Irish Cancer Society, provided just such an opportunity. The event took place at the Shelbourne Hotel – within walking distance of Dáil Éireann – on the Wednesday before the Fianna Fáil ard-fheis.

The hospitality-industry interests had dominated the lobbying platform for weeks. They had achieved a great deal of coverage and had won further public support. By this stage, however, they had run out of time and opportunities. It was now time for the ban supporters to have their say; the manner in which Jane Curtin orchestrated the response of the ban supporters was highly effective.

The venue booked by Jane Curtin was the famous Constitution Room in the Shelbourne: eighteen different organisations, all of which supported the ban, gathered to demonstrate the benefits of the smoking ban for each of the sectors they represented. Collectively, these organisations represented some 1.1 million members, and each group had spokespersons at the meeting to articulate their positions of support for the ban. The very fact of all of these organisations joining forces for a collective public demonstration of support was in itself very significant. It served to galvanise the many interest groups: heretofore, they had been working effectively behind the scenes, but they were now given the chance to play a role in a very visible display of solidarity.

Every member of the Dáil and Senate was invited to make the short stroll from Leinster House for the meeting. Jane Curtin also pointed out to the deputies and senators that the national media would be present. It was a case of 'stand up and be counted' as far as the politicians were concerned: some sixty-six parliamentarians attended the event.

The Constitution Room proved to be far too small for the number of people who attended. After the organisers of the event had ushered him through the crush of health professionals and politicians, Micheál Martin was met by a throng of media – more befitting a Budget-day press conference – that was anxious to know whether, with only three days to go to the ard-fheis, he was having any second thoughts on the exact form of implementation of the ban. Martin addressed the large crowd and announced that he was now at the stage of preparing to sign the all-important regulations as soon as the final draft was complete, in a matter of days.

Professor Luke Clancy also addressed the meeting. He reiterated the message that 'Environmental tobacco smoke kills. This legislation offers a special once-off incentive to those who smoke to quit, and safety to those who do not.' John Douglas, national official of the bar workers' union Mandate, issued the following statement: 'The Minister for Health's proposed ban on workplace smoking has the unequivocal support of Fine Gael, Labour, Progressive Democrats, Greens, Sinn Féin and a number of the independents. The ball now rests in Fianna Fáil's court, and their deputies now have to decide where they stand. They have to show whether they are committed to protecting workers from the ravages of cancer and heart disease, or whether they are going to succumb to the lobby of an interest group whose case is based upon unfounded allegations of potential economic damage.'

The exposure of so many politicians to such a unified demonstration of solidarity and support for the smoking ban had a profound impact. It also served to prove in a uniquely powerful manner to the media present that, despite all the rhetoric and activity from the opposing side, the people who had believed from the outset in the efficacy of the ban were more convinced than ever that putting the saving of lives above profits was the only way to go.

The representative groups who participated in the public demonstration of support for the ban were the Irish Cancer Society, the Irish Heart Foundation, ASH Ireland, the Irish College of General Practitioners, the Irish Hospital Consultants Association, the Irish Medical Organisation, the Irish Nurses Organisation, ICTU, the Alpha One Foundation, IMPACT, Mandate, SIPTU, the Asthma Society of Ireland, the Environmental Health Officers Association, the Institute of Public Health, the Irish Sudden Infant Death Association, the National Heart Alliance and the Irish Countrywomen's Association.

*

Rachel Sherry, the PR consultant working on behalf of the Health Promotion Unit, had identified an anti-smoking expert who was ideally placed to comment meaningfully, from first-hand experience, on the impact of smoking bans elsewhere and the difficulties that had been encountered in implementing such bans. JoAnn Landers, director of the tobacco control programme for the city of Boston, had overseen the introduction there of a blanket workplace smoking ban in May 2003.

Landers travelled to Ireland to discuss various aspects of the implementation of the proposed Irish ban and to comment on her recent experiences, especially in dealing with the hospitality industry in Boston. The reactions that her office had encountered from the bar and hotel sectors in advance of the Boston ban were similar to those experienced by the Department of Health, with owners expressing concern about loss of business and jobs. Landers said that the Boston Health Department had taken time to educate both the trade and the general public about the dangers of ETS and had built up public acceptance of the ban to 80 percent by the time the ban was introduced. Her office had also worked with the bar trade and had built up a 97 percent rate of compliance with the ban.

'My message to the Irish publicans is that there has been no loss of business in Boston,' she said. 'We have over sixty Irish bars, and not one in the group sample survey has lost an employee. Business is booming. People are coming out, especially those who never went to bars before, because they can breathe clean air when they are drinking, eating and socialising.' Of 650 licensed premises inspected during the first months of the ban in Boston, twenty had been fined for being in breach of the smoking ban, Landers said. She went on to comment that the hospitality industry in Massachusetts had tried to reach a compromise with the city, involving better ventilation, but their arguments could not be made stand up, with study after study showing that ventilation was ineffective in reducing levels of air pollutants associated with ETS.

In an interview in the *Irish Independent*, Landers described the arguments put forward that the ban was unenforceable and would lead to excessive noise and unruly behaviour as 'a ploy on the part of the tobacco industry'. She continued: 'Publicans and those in the hospitality industry are getting pressure from the tobacco industry. Fears are being instilled in them that people will be unruly and the ban will be unenforceable. They are being lied to. The tobacco industry has a lot to lose if this ban is passed.'

She added that similar tactics to those being adopted by opponents of the ban had been noted by the health authorities in Boston in advance of their ban. 'All of a sudden, a lot of alliances popped up, and the same arguments about noise and litter were used. In Ireland, it sounds a lot like Big Tobacco to me, although I'm not able to point the finger directly at them.'

*

The focus of the latest phase of the publican-led campaign against the ban had been to exert as much pressure as possible on delegates heading to the Fianna Fáil ard-fheis. The anti-ban PR machine was still in top gear as delegates arrived in Killarney to attend the conference, which was being compressed into a one-day event as distinct from the traditional two-day affair. One of the founders of the IHIA, Paudie O'Callaghan, who had been a member of the original delegation that had visited New York, decided to highlight his objections to the ban by enlisting his own 'smoking police' to impose a temporary ban on smoking in the town's pubs during the visit of the Fianna Fáil entourage. O'Callaghan arranged for a team dressed as police officers to issue 'tickets' to anyone found smoking in a bar in Killarney during the ard-fheis. He said that the exercise was designed to show how unworkable the ban would be. As chairman of the VFI in Killarney, an organisation that represented 360 publicans, he said that they had decided not to protest formally at the ard-fheis but instead to wait until January. 'We've exhausted the political route,' he explained, 'so now we're just going to refuse to implement the ban.'

How much impact did all this PR and lobbying activity have on the delegates? As they gathered at the Gleneagle Hotel in Killarney, *Irish Times* correspondent Lorna Siggins submitted a story based on comments by anti-ban minister John Browne. He was quoted as saying that 'The Ministerial Order will be signed next week and there isn't a damn thing we can do about it. I just hope that it will be implemented in a reasonable manner and that we won't see individuals and publicans going to jail over it.' So the huffing and puffing was coming to an end.

Instead of the usual format of motions and debates, in 2003 the ard-fheis was, for the first time, based on ministers setting out their programmes and delegates having the opportunity to pose questions in workshop settings. In his keynote address, the Taoiseach, Bertie Ahern, bluntly announced that the smoking ban would go ahead, with no compromise. This drew the loudest cheer of the whole ard-fheis. Despite all the previously reported controversy and apparent divisions within the party on the issue, when it came to the crunch, party loyalty prevailed and dissension evaporated. Minister Martin was the undoubted hero of the 2003 ard-fheis. A report by Harry McGee of the *Irish Examiner* summed up the outcome for Martin: 'With one eye-catching policy, Micheál Martin seems to have pulled off a Houdini-like escape from the multifarious woes that dog his problematic portfolio.'

The ard-fheis's endorsement of the smoking ban was overwhelming. To loud applause, Micheál Martin had restated his commitment to the ban: 'We are going to bring in the regulation,' he said. 'I am going to sign the legislation.' Martin was later to refer to the reception he received on his arrival at Killarney as one of the highlights of his lengthy campaign in the lead-up to the introduction of the ban. Before he had made his way to Killarney, the last question he had been publicly asked, live on air by Pat Kenny, was: 'Are you prepared to be eaten alive in Killarney by your party colleagues?'

In fact, quite the opposite was the case. Martin recalls: 'When I arrived in the lobby of the hotel, the delegates started queuing up to shake my hand. Everybody was saying: "We're all behind you. Keep at it. We're backing the smoking ban, and don't let the critics get you down." It wasn't orchestrated. Some people might say that we had wheeled out the grass roots. We didn't. In fact, we did not know what way the smoking ban would play with the backbenchers. What happened was that, after the formal session, I could not get out of the hall because of the number of people lining up to say just one thing: "Keep it up; we believe in the smoking ban.'"

This account is corroborated by Mark Brennock of the *Irish Times*, who, reporting on the ard-fheis, wrote that 'the fanciful notion that a revolt on the smoking ban would become apparent was quickly shown to be the wishful thinking of ban opponents. Mr Martin repeated that there would be no compromise and got substantial applause. This issue, long portrayed in some quarters as the harbinger of a "grass-roots revolt", appears well and truly over.'

The Taoiseach concluded his ard-fheis comments on the issue by turning his attention to the licensed trade. He urged publicans to put the health of their employees and the public first. 'The forthcoming legislation is the most impressive and enlightened I have come across,' he stated, 'and I urge that there will be no "halfway" compromise.'

In an interesting postscript to the ard-fheis, the *Sunday Tribune* published the results of a poll, conducted on its behalf by Millward Brown IMS, that assessed the popularity of politicians and focused on possible leading candidates to succeed Bertie Ahern as Taoiseach. It was a major surprise to many that the leading contender by a substantial margin was Micheál Martin, who, as minister for health, had been at the receiving end of negative publicity on various issues for almost four years.

So on Saturday 11 October, Martin successfully negotiated a major hurdle in his campaign, by winning a standing ovation and the undisputed support of his own political party. Judging by the depth of feeling he had encountered during his appearance on *The Late late Show* on the Friday night before he travelled to Killarney, however, although one battle – the one within Fianna Fáil – had been won, the war against the ban was still raging.

12

The *Late Late* Showdown

Nothing is a serious issue in Ireland if it is not teased out, debated, challenged and turned into a verbal boxing match on *The Late Late Show*. For decades, the stock-in-trade of television supremo Gay Byrne was the phenomenal influence his TV show had on the national lifestyle. This legacy was carried forward whenever the opportunity arose by Byrne's successor, Pat Kenny, but heading into the new autumn TV schedule in 2003, a new phenomenon was to emerge onto Ireland's chat-show scene: *The Dunphy Show* on TV3.

For months prior to the rival shows hitting the airwaves in the new season, the competition between *The Late Late Show*, the chat show with the world record for longevity, and the show headed by the unpredictable, often outrageous, always controversial Eamon Dunphy had been hyped for all it was worth by the media. The producers of both shows promised the biggest star line-ups and the liveliest debates on the hottest topics of the day. So an interesting conundrum faced Rachel Sherry and the pro-ban PR team in choosing which show would be best for the showdown debate on the smoking-ban topic – a subject that would inevitably exercise the attention of top TV producers.

Dunphy would undoubtedly attract a large initial TV audience that was curious to see how well he could deliver on his promise to trot out the biggest stars and also to see whether – as seemed likely, given his broadcasting reputation – he would come up with anything to shock his audience. On the other hand, *The Late Late Show* would be under pressure to preserve its premier position in the history of Irish television. The producers would not surrender their audience without the greatest struggle, and on such a serious issue it was not a time for the Department of Health to gamble or to seek controvery. Rachel Sherry opted for *The Late Late Show* and set about trying to ensure balance in the debate, in return for delivering to them an exclusive debate on the topic of the moment.

'Getting on the *Late Late*' is one of the main targets of every PR person in Ireland who aims to promote an idea, product, event, book or any form of entertainment. A well-prepared presentation on the slickest show on Irish TV, delivering a vast audience, is a sure-fire method of getting one's message across. But the show is a two-edged sword: if a guest gets it wrong, the results can be catastrophic. The list of guests who have 'blown it' on *The Late Late Show* is long in number and impressive in character. Former EU Commissioner Padraig Flynn was regarded as having lost the plot when listing the various perks and financial benefits of holding the job of Commissioner. RTÉ's own Andy O'Mahony never recovered from the shocked public reaction to his question to Gay Byrne's biographer, Deirdre Purcell, about whether she was sexually attracted to Gay – a jokey comment that went horribly wrong.

Micheál Martin would certainly run a risk by appearing on the show: he might allow himself to be distracted into sticking to the agenda of the ban's opponents; alternatively, he might appear on camera to be lacking in assertiveness or commitment. Nonetheless, he was secure in his belief that the proposed ban on smoking in the workplace was a very credible initiative that offered genuine benefits in terms of public health, and he could be relied on absolutely to stay 'on message'. His advisers were fully behind his personal involvement in what would undoubtedly be a heated debate. On the other hand, the hospitality sector had come up with a litany of arguments against the ban and was convinced that its side of the story was winning increasing public support. Here was a golden opportunity to put the minister on the spot in order to restate categorically his conviction regarding the ban, while at the same time facing down the many interest groups on the hospitality side who would use the occasion to vent their anger at the ban.

Rachel Sherry had a very hectic and sometimes frustrating week in the lead-up to the show. Knowing that the aim of the broadcasters would be to encourage an all-out attack on the minister – the reputation of the programme was based on its providing a platform for the most heated arguments ever broadcast on the airwaves – it was important to ensure that the representation of pro- and anti-ban campaigners would be as balanced as possible. Finding out how many places would be allocated on the panel was a major task in itself; securing places in the audience for supporters of the ban was another matter. It appeared that publicans from every corner of Ireland were pulling strings to obtain seats in the restricted audience, so spaces were at a premium and it would be difficult to accommodate many of the experts who could contribute evidence or views to the debate.

One expert whom, Sherry felt, would be a particularly important contributor was JoAnn Landers. As the hospitality side would no doubt include a New York publican in its line-up, it was decided that it was important to have present an authoritative spokesperson with first-hand knowledge of the effectiveness or otherwise of the bans in the US. Despite every effort to have Landers included as a member of the panel, it transpired that she would only be accommodated in the general audience; the programme team at least gave a commitment that Landers would be given an opportunity to speak, however. Of course, similar undertakings were undoubtedly given to the many other notable guests, who represented so many parties who were concerned about the smoking issue. In the event, the main panel consisted of four guests: on the 'pro' side, Minister Martin and Dr Shane Allwright, who had written the initial report recommending the all-out ban and was described by Pat Kenny as 'the lady who is responsible for all this', and on the 'anti' side, Tadg O'Sullivan of the Vintners Federation of Ireland and Finbar Murphy of the Irish Hospitality Industry Alliance.

Straightaway it was clear that some interesting choices had been imposed on both sides of the debate. For the pro-ban team, the choice was relatively straightforward. The minister was in the hot seat. He would be expected to field all questions from every quarter, but he would have an authoritative expert on his side. In addition, Professor Luke Clancy, the leading respiratory consultant and one of the most outspoken supporters of the ban, was in attendance, as was Dr Fenton Howell, a very articulate anti-smoking campaigner. Norma Cronin of the Irish Cancer Society – and the society's specialist on smoking cessation – was also present, as was Maureen Mulvihill of the Irish Heart Foundation, Tom Beegan of the Health and Safety Authority and representatives of Mandate.

Probably the most significant decision in relation to the speaker panel was left to the hospitality side. From the outset, the three main opponents of the ban had been the Irish Hospitality Industry Alliance, the Licensed Vintners Association and the Vintners Federation of Ireland. In light of the rumblings of discontent amongst publicans regarding their organisations' failure at times to deliver the clout the pub owners felt they deserved, it was interesting that the powerful, Dublin-publican organisation the LVA was sidelined. So it fell to Tadg O'Sullivan of the VFI, who had the most live broadcasting experience of anybody in the Irish pub trade, and Finbar Murphy, of the relatively recently founded IHIA, to fight the battle on behalf of publicans and hoteliers.

Clearly the line of attack from the ban opponents had been tactically agreed in advance. The confrontation with the minister would be aggressive. It would aim to portray him as intransigent, out of touch and more concerned with his own political future than with the damage his new law would cause. The opponents of the ban would emphasise the impact of the measure on jobs and the fact that their reasonable proposals for compromise were not being listened to. They would also attempt to challenge the need for a total ban, when they were prepared to invest in ventilation systems which they believed would adequately deal with the problem of environmental tobacco smoke.

All of these tactics and issues had been predicted by the minister's team of advisers. His press secretary, Caitriona Meehan, had been in a series of intensive meetings with Rachel Sherry and her team in the days leading up to the Friday show. Inevitably that week, the minister's schedule was as hectic as ever. Of all the ministerial portfolios, the health ministry is one of the most gruelling in terms of briefings, meetings and attendance at events, as well as the normal business of managing one of the top departments in terms of government spending. The problem facing the team was to find a time-slot when the minister could sit down with them to go through a mock question-and-answer session, which is so important in preparing for a live TV debate. Indeed, a suitable time-slot did not become available until late Friday afternoon, just hours before the broadcast.

Micheál Martin sat down with his policy advisers, Deirdre Gillane and Christy Mannion; representatives of the Office of Tobacco Control and the Health and Safety Authority; Noel Usher of the Department of Health and the team that was drafting the regulations and guidelines for the new law; Chris Fitzgerald of the Health Promotion Unit; his press officer, Caitriona Meehan; and Rachel Sherry, who was providing up-to-date briefings on the activities and status of the non-government organisations. Everybody in attendance realised that the TV debate would be an important step, not from the point of view of influencing the decision to proceed, but in terms of galvanising the many supporters of the ban in their resolve over the remaining months leading up to implementation of the ban and in reassuring the public at large that the ban was all about saving lives and providing a healthier environment for workers and the general public for generations to come. No doubt the show would also be watched by Martin's party colleagues gathering for the following day's congress in Killarney.

The final details of the regulations, which would soon come into law, were still at drafting stage, so the minister needed to be brought up to speed on these important practical points. Martin was also briefed on the most recent research findings on ventilation and the latest opinion polls on the ban. He knew precisely where the hospitality sector would be coming from and the main points of their latest arguments, having met a delegation of their representatives earlier that week. He now needed to place all of these arguments in context and be comfortable in his own mind that, despite the aggression he was about to face, this live TV debate would be an important opportunity within the overall information campaign in support of the ban. The minister would therefore focus on the core message that the ban was in the best interests of protecting workers in every workplace in Ireland, and he would reaffirm his commitment to the ban. He had expert advice to prove it was a necessary step, in view of the 7,000 lives lost in Ireland to cancer each year and the continuing high incidence of heart disease in Ireland.

The major challenge for the minister was to remain focused on his messages in the face of stern opposition from a panel of opponents who were very agitated indeed. His other challenge was to attempt to persuade the hospitality sector to face up to the inevitable and to work towards making the ban not only workable but, where possible, beneficial to them.

Out at the RTÉ studios, the main preoccupation for the TV station was the battle for that night's audience. The upstart chat-show presenter on TV3, Eamon Dunphy, kicked off that evening with a scoop: he had convinced Roy Keane, Ireland's most controversial soccer star, to undertake his first live TV interview since his infamous departure from the World Cup squad. Dunphy had also lined up Tony Blair's former communications adviser, Alastair Campbell, to try to justify Britain's involvement in the war in Iraq. Also on the guest list was former Boomtown Rat Sir Bob Geldof. Pat Kenny's line-up was equally impressive that evening, with George Best and his former Manchester United colleague Denis Law, and Daniel O'Donnell performing his new single. Most important, the first live TV confrontation on the controversial smoking-ban issue assured *The Late Late Show* of a high-impact start to the show, which was likely to win it the biggest audience share – which indeed it did.

As well as the pro-health representatives in the audience, Pat Kenny had a very wide selection of ban opponents to choose from, in order to ensure the vital ingredient for a successful TV chat show – a lively verbal row. Very quickly, the markers were laid down by the speakers' panel. The minister was not for turning, and he speedily set out his stall once again on the purpose and justification for the ban. Probed by Pat Kenny on whether a total, blanket ban was feasible, the minister alluded to the regulations being drafted, which would refine the implications in relation to 'home' situations, such as mental institutions and prisons.

The opening foray by Tadg O'Sullivan on behalf of the publicans was, in the opinion of many in the communications community, a tactical mistake. The representative of the pub trade was present for the purpose of professionally outlining the impact of the ban on the operations of the publicans and challenging the ban on the basis that it was an unnecessary intrusion on the commercial welfare of the pub sector throughout the country. Instead, he chose to make opening comments condemning the general situation regarding the health services in Ireland. His line of attack was to repeat the suggestion that the smoking ban was a smokescreen to divert attention from the general problems in the health service. The golden rule for spokespersons is to confine their comments to the areas in which they will be recognised to have expertise. For a spokesperson with specific commercial interests to attempt to make capital by offering observations on wider social, economic or polit- ical issues or on matters outside their realm of business is a sure- fire way for them to lose the sympathy of an audience. O'Sullivan, however, later made the point forcefully that the minister had not been listening to the arguments of the publicans, who were now particularly agitated because their representations appeared to have been ignored. In O'Sullivan's opinion, ventilation was the answer: publicans should have the option of investing in enhanced ventila- tion systems or going out of business, he declared. When pressed on several occasions by Pat Kenny, however, O'Sullivan could not come up with an estimate of the cost of a suitable ventilation system.

With cigarette-vending-machines operators, ventilation-system manufacturers, publicans, guesthouse operators, hoteliers and a large selection of publicans to choose from – all of whom were prepared to state that the ban would result in job losses – Pat Kenny took most people by surprise by choosing to focus on bingo oper- ators for the first challenge to the minister from the general studio audience. Siobhan Kearney had incurred a fine earlier in the week for her bingo-hall customers having failed to observe a no-smoking ban. She and a colleague outlined the difficulty of policing and enforcing a ban, even amongst the bingo hall, middle-aged to eld- erly, mainly female, clientele, many of whom flatly refused to observe the smoking ban. In typical journalistic fashion, the *Late Late* team had come up with one of the more unpredictable angles in an effort to throw the panellists, notably the minister, off guard. His response that, in his youth, he had worked in smoky bingo halls and that there was 90 percent compliance with existing smoking bans was followed by the theme swiftly changing to the impact of the ban on jobs. This prompted Finbar Murphy of the IHIA, who had obviously been well groomed by his professional advisers for the occasion, to stress the impracticability of aspects of the proposed legislation, especially in regard to policing what went on in bedrooms of hotels, some of which could have 300 to 400 rooms. Murphy made his points clearly and succinctly and sought to prove that the ban would be unworkable.

At this point, Pat Kenny invited comments from the pro-ban side: the US visitor JoAnn Landers was first to speak, stating that, since an identical ban had been introduced in Boston the previous May, the city had achieved 97 percent compliance and the enforcers of the new law were working closely with the business community to achieve success based on mutual collaboration. Landers equated the need for the introduction of a smoke-free environment with measures to protect workers against asbestos. She also made the point that, in Boston, the implementation of the ban had to a large extent been self-policing, with non-smokers encouraging others not to smoke. Professor Luke Clancy made the point that it was disappointing that publicans seemed to be more concerned with profits than with protecting the health of their employees. Dr Fenton Howell added that the concern among publicans that customers would revolt and cause problems was groundless. He said he believed that smokers would be no less law-abiding than non-smokers.

The discussion certainly proved to be one of the more spontaneous and lively television debates on current affairs for a very long time. Pat Kenny had ample contributors to choose from and ensured that as many as possible got the chance to air their views. By the end of the programme, there were no winners or losers in terms of point-scoring, and no views or opinions amongst the protagonists had been changed. Probably the most beneficial outcome of the debate for all concerned was the fact that they had expressed their opinions to the largest public audience in Ireland.

The last word was offered to Minister Martin by Pat Kenny, who opined that the minister was heading off to Killarney to 'get it in the neck from the Fianna Fáil backbenchers.' Martin replied that he was not worried about 'popularity stakes' and that he believed he had no choice but to implement the ban. 'This is a very important health decision,' he said. 'I'm happy I am doing the right thing for the Irish people. I'm also satisfied that generations to come will look back and say this was the right decision.'

Who won the TV-ratings battle on the night? Despite Eamon Dunphy's wildcat reputation and his appeal to a certain clique of celebrities, *The Dunphy Show* on TV3 came off second-best. The decision to have the live debate on the smoking ban on *The Late Late Show* was the correct one. Incidentally, *The Dunphy Show* did not last until Christmas 2003.

13

The Health Campaign Moves Up a Gear

With the major political hurdle of the ard-fheis out of the way and final details of the new legislation being drafted for publication in October, it was time for the ban promoters to move forward from what had essentially been a monitoring role and to press the green light with their proactive campaign plans. These plans were very comprehensive and were geared to enlisting a wide-ranging alliance of health professionals in the implementation of the ban.

From the outset, the establishment of a central press office was undertaken as a priority. Cohesion and consistency in the flow of information was seen to be of paramount importance, and Rachel Sherry was assigned the role of coordinating information on all of the initiatives being undertaken by the various organisations in the health alliance. This also facilitated the management of media referrals and queries to the most relevant spokespersons in all eleven health-board areas. Such an approach proved to be a very efficient way of capitalising on the high level of cooperation that existed amongst the pro-health campaigners and of maximising the expertise that they made available to the communications team.

General practitioners would play a key role in the campaign for the ban: one of the first proactive tasks of the pro campaign was to highlight the public endorsement of the smoking ban by doctors throughout the country. Dr Prannie Rhatigan, director of the smoking-cessation programme of the Irish College of General Practitioners (ICGP) and a member of the steering group on the ban, issued a statement and mobilised the nationwide doctor-spokesperson network to push the message that doctors were four-square behind the ban and 'strongly feel there is no room for compromise.'

Dr Richard Curran, chairman of the ICGP, issued a call to GPs to urge employers, employees and the general public to embrace the ban. Rachel Sherry and her team set about highlighting case histories featuring former bar workers and entertainers who were obliged to work in smoky environments and also sought out examples of workplaces that had already introduced smoke-free policies voluntarily. Unions, Chambers of Commerce and health-and-safety officers were approached to participate in the initiative.

Ireland's top TV soap, *Fair City,* was one of the first popular media platforms to feature the ban, highlighting that it was being introduced at some of the country's best-known hostelries, including McCoys Bar, the Blue Dolphin and the Sandwich Bar. The same applied to the TG4 soap *Ros na Rún.*

Health boards around the country set about organising localised media and community-based publicity initiatives. Local sports heroes were enlisted to demonstrate their support for the ban and to urge people to quit smoking. Indeed, quitting was a very important complementary message to the awareness campaign about the workplace ban. The fact that the Health Promotion Unit had a mandate for two separate information initiatives, one to generate awareness of details of the impending ban and the other to motivate and support smokers who wanted to quit, was a great benefit. A national smokers' 'Quitline' had been set up in conjunction with the Irish Cancer Society; this facility was promoted extensively throughout the health-board network and the health community countrywide. The Quitline was operated seven days a week, from 8 AM to 10 PM, and drew an average of 700 calls per day.

The promotion of the Quitline coincided with a hard-hitting advertising campaign undertaken by the Health Promotion Unit, which had the strap-line 'Every cigarette is doing you damage.' Chris Fitzgerald of the Health Promotion Unit was responsible for the utilisation of what were acknowledged to be two of the most graphically impactful commercials ever seen on Irish television. 'We had several creative options available to us but the dramatic visual representation of the damage caused to the brain and to the arteries was the option which researched most positively, especially with young people,' Fitzgerald commented. 'There is always a danger that TV viewers will reach for the zapper when they are confronted by images that shock or horrify, but in this instance the impact was very strong and viewers took on board a very powerful message about the damage caused by cigarette smoking. The minister shared our opinion, when it came to deciding that shock tactics were the best option. He was very much up for it, and it has proven to be the best strategy for motivating people to quit.'

The Pharmacy and Dental Healthcare sectors were also brought into the pro-ban network, both to support the advice to quit that was being promoted by doctors and to advise on nicotine-replacement therapy (NRT). A major distribution of smoking cessation intervention packs was achieved through this network. The fact that GPs could prescribe free NRT products was important, as was the fact that a wide range of therapies, including patches, gum, lozenges, tablets and inhalers, was available over the counter. Unprecedented sales of these products resulted from the concerted emphasis during this period on the dangers of smoking.

A 'Smoke-Free at Work Advice Kit' was drafted by the Irish Heart Foundation and the Irish Cancer Society and developed by the Grayling agency in conjunction with the Tobacco Control Unit of the Department of Health and Children. This kit aimed to facilitate employers on how best to prepare for the implementation of the ban. Employers were, naturally, a key target for the campaign organisers, and a mass mail-out of materials was arranged, including leaflets summarising the main details of the ban, posters announcing that the ban was about to be introduced, and 'no smoking' items such as beer mats and free-standing table cards for the licensed trade. This was in addition to the formal guidelines that would later be issued by the Office of Tobacco Control and the Health and Safety Authority explaining the legislative implications of the ban and the responsibilities with regard to its implementation for those in charge of workplaces. The information programme for employers was supported by a special 'Smoke-Free at Work' website.

A bank of research findings and national and international

statistics had been built up earlier in the year; this was now made available to the spokesperson network and to the appropriate databases that had been established. This was a very valuable information resource, especially when it came to ensuring that all proponents of the ban were singing off the same hymn-sheet.

Meanwhile, conscious that the ventilation issue was likely to surface time and again, as it appeared to be one of the main platforms of the VFI counter-argument to the ban, it was decided that Maurice Mulcahy of the Western Health Board, who had extensive knowledge in this area and who had established a strong line of communication with one of the best-known international experts on ventilation, Dr Jim Repace of the US, would be deployed to explore every aspect of this issue. Mulcahy duly joined the special steering group, which was operating under the aegis of the Health Promotion Unit.

Another important consideration was for all pro-ban campaigners to function under a common public identity; it was therefore agreed to 'brand' the programme elements under a common theme. From a wide range of options considered by the steering group, 'Smoke-Free at Work' was finally decided upon as representing the thrust of the combined efforts. This was adopted as the headline for all literature and promotional items to be used in the campaign.

On 23 October, Micheál Martin announced that he had signed the Tobacco Smoking (Prohibition) Regulations 2003, made under the Public Health (Tobacco) Act 2002. The new regulations on smoking in the workplace were announced as coming into operation on 26 January 2004, almost one year after the initial announcement of the measure. The minister announced that the new regulations would apply to all enclosed places of work other than private dwellings. He also announced that the unique circumstances pertaining to prisons and places of detention were the subject of ongoing consultation between the relevant agencies and that he was awaiting the advice on this issue of the Office of the Attorney General. The matter of prisons and psychiatric units being designated places of work, but possibly qualifying as exemptions, would be dealt with by an implementation committee that consisted of senior Department of Health officials, the Office of Tobacco

Control, the Health and Safety Authority and other relevant institutions. The committee would be chaired by Noel Usher, of the Department of Health.

The minister used the occasion of this announcement to restate that he had been advised that current ventilation technology was ineffective in removing the risk of passive smoke to health and that the problem of environmental tobacco smoke required legislative measures. Explaining the selection of 26 January as the date for the implementation of the ban, the minister referred to the need to synchronise the date with the statutory period required by the European Union from the date of publication of the regulations.

Prior to the announcement, there had been considerable discussion amongst the communications team about the format of the announcement. Since the ard-fheis, the media had been anticipating the announcement, and they could well treat it as a formality. So would a statement or press release suffice? Or would a routine briefing of the health correspondents, with whom Caitriona Meehan was in daily contact, be enough? A full-blown press conference might be seen as being over-the-top for a topic that had been debated so widely to date. In the event, the minister himself made the decision to announce the regulations at a full media briefing in the Media Centre at Government Buildings. This transpired to be a wise selection, with the venue packed to standing room only, with news reporters, health, industry and political correspondents, and many from international media organisations present.

Micheál Martin concluded his announcement with an appeal to vintners to reflect on the reason behind the new legislation and to work towards its implementation. He went on to warn publicans that they could leave themselves open to legal action from employees if they did not provide the protection afforded by the ban. Questioned about possible difficulties in policing the ban and securing compliance, especially in pubs, the minister responded that a 'significant number' of environmental-health officers would be deployed; in relation to compliance, he expected that, as was the case in other jurisdictions, public support for the measure would help to ensure that it was policed effectively.

In general, the national media seemed almost relieved that the new legislation was now official – at last. The *Irish Independent,* in an editorial the next day, commented that 'after all the huffing and puffing', the minister could breathe a sigh of relief that, after a hard-fought campaign, his new legislation was now in place. Nonetheless, the minister still faced the challenge of the successful implementation and policing of the ban.

As it transpired, this was not the only challenge the minister would face. Within hours of his media briefing, the IHIA announced that it would be referring the new regulations to its legal advisers to seek the possibility of a judicial review. Finbar Murphy of the IHIA also seized on the lack of clarity concerning prisons and psychiatric institutions, claiming that the possibility of exempting prisons and any other forms of accommodation made a mockery of the claim that the ban was to protect the health of all workers. 'To make exemptions is an acknowledgement that the proposals are unimplementable,' Murphy stated. 'Either the policy is zero tolerance or not, and the minister needs to clarify this.'

Martin responded by asking why the lobby group would consider taking legal action against a measure designed to protect and improve public health. He also said he was on firm legal ground in relation to prisons and places of detention because these locations were already covered by special provisions under the 1989 health and safety legislation. In the final move in this particular episode in the smoking-ban saga, the *Sunday Times* later that week reported on a nationwide fund-raising drive being undertaken by publicans and hoteliers throughout the country to finance a legal challenge to the ban.

So, having faced down the opposition within his own party and introduced the critical piece of legislation, Micheál Martin had successfully navigated around a crucial stage in the process. Now, within a matter of days, however, he found himself confronted with a whole new set of challenges.

14

The Publicans' Anger Boils Over

'There is no way I'll be enforcing it, even if the minister himself comes in,' threatened Jackie Healy-Rae, the Kerry publican and TD. The colourful independent deputy was milking the media coverage on the smoking ban for all it was worth and was establishing himself as one of the most outspoken opponents of the ban. In fact, his actions bordered on the outrageous when a photograph of him in a convict's outfit appeared in national and local newspapers. Yes, Healy-Rae wanted to be seen as being ready to serve time rather than refuse to serve anybody who was intent on smoking in his pub. He was joined in the photographs by fellow publicans John O'Sullivan and Danny Leane of the Kerry Vintners Association, both also in prison garb and handcuffed.

This was a light-hearted, farcical episode which no doubt delighted newspaper picture editors, but there was certainly nothing light-hearted about the mood of publicans countrywide when they realised that, with the signing and impending implementation of the new smoke-free regulations, the threat of the smoking ban had become a reality.

Indeed, when he was not posing for silly photographs, John O'Sullivan was a leading light in his local association. Kerry publicans were the first to commit publicly to a decision to defy the ban by refusing to implement it when it came into effect. Following a meeting of 200 publicans, who voted overwhelmingly against the ban, O'Sullivan referred to the situation which he claimed had prevailed in Holland, where a smoking ban which was meant to have been introduced in January 2004 was deferred for two years to facilitate consideration of other options put forward by the hospitality and licensed trades. He told the meeting: 'I have seen a sea change in public opinion over this issue.'

Whereas, heretofore, the publicans' mood had been somewhat militant, by the latter part of October it had become very angry indeed. Although the Kerry publicans were among the most outspoken in their opposition to the ban, it was clear from a series of very large meetings of publicans held around the country that the anger was spreading nationwide like bushfire.

'Bad law will always be defied,' said the VFI, although they were careful to declare publicly that it would be a matter for individual publicans to make up their own minds about whether to implement the ban or not: the VFI would facilitate meetings and opportunities for publicans to vent their anger on the issue, but the federation would not urge their members to break the law. But 'jails full of publicans' was to become a common threat emanating from the many publican meetings.

In Cork, 950 pub owners and managers expressed their determination to fight the ban to the bitter end. A vote of members of the VFI resulted in unanimous agreement on refusing to observe the new law. Despite receiving confirmation from the Department of Justice that failure to comply with the ban could lead to objections to their pub licences being renewed, the Cork publicans, through their press-relations officer, Con Dennehy, vowed to appeal any challenges to their licences by referring their cases to the highest courts available. Tadg O'Sullivan of the VFI confirmed to those present that the best possible legal advice would be made available on this issue. The Cork publicans were also told that a fund was being established to cover the costs of fines incurred by members who were found to be in breach of the law.

In Waterford, a similar outcome was reported from a meeting of fifty publicans. 'We don't want to end up in court or in jail but we are prepared to face the consequences because we have no other choice,' said Michael Fitzgerald, the chairman of the Waterford Vintners Federation. The Waterford publicans voted unanimously not to implement the ban.

As well as stating their determination to break the law rather than ask their customers to put out their cigarettes, publicans began to propose the idea of running their own candidates in future local and general elections. This was claimed by the VFI as the most acceptable means of restoring democratic representation for publicans and their customers, on the basis that the existing public representatives had so far failed to deliver what the publicans were demanding.

As the campaign of publican meetings gathered momentum, 302 members of the Clare branch of the VFI unanimously voted to defy the ban. The meeting was influenced by a report from chairman Declan Brandon of a former Clare publican now based in Florida who had experienced a 45 percent drop in his business as a result of a partial ban introduced in the US state.

Publicans in Ballybofey in County Donegal also came out strongly against the ban. Their meeting, which was attended by more than 100 of the 350 VFI members in the county, aimed to highlight the fact that, throughout their area, it was mainly family-run pubs that would be affected by the ban.

Sligo, Monaghan, Louth, Leitrim, Cavan and Mayo witnessed similar highly agitated anti-ban meetings as feelings became more strident among the licensed-trade community. In Bailieborough, publican Pat Langley, addressing a meeting of Cavan pub proprietors, said: 'There's concrete evidence of functions being cancelled the length and breadth of the border and the business going across the north to premises where people will be able to smoke in peace.' This message was typical of some rather curious and highly questionable observations being made at border-county publican meetings months before the ban was due to be introduced.

'Whether we are legally entitled to or not, we won't be implementing the smoking ban,' declared Marty O'Hora of the Mayo VFI as resistance to the ban continued to spread. Elsewhere in the west, Galway publicans were downbeat, but took heart from the decision by publican Val Hanley to resign as chair of the Western Health Board as a protest against the ban. The meeting agreed to continue to exert pressure on local minister of state Frank Fahey, who was known to be grappling with some tricky issues regarding difficulties with existing health and safety legislation and its appropriateness to the implementation of the new law.'

Fahey's fellow minister of state, Brian Lenihan TD, came in for a great deal of criticism when he arrived to celebrate the re-launch of the Balbriggan Chamber of Commerce. What was billed as a pleasant official visit turned out to be something quite different, with the ministerial car passing down the main street, which was lined with picket-carrying publicans. Indeed, while the Chamber president was delivering his welcome speech to the minister, the Chamber vice-president, Gary Browne, the owner of DeBrun's pub in the town, was outside marshalling the protestors. Patrick McCormack, owner of the White Hart Inn in Balbriggan, told the minister that he had been behind the counter since the age of four, had started washing bottles at the age of six and had worked behind the bar 'when you wouldn't recognise a customer at the other end of the bar, there was so much smoke. No one who ever worked there died from cigarettes,' he declared.

With county meetings being held in quick succession in Tipperary, Limerick, Wexford, Westmeath, Laois and Kildare – and all fuelling the rising tide of protest – the scene was set for a massive show of strength when the VFI convened a national meeting at the Heritage Hotel in Portlaoise and urged all of its 6,000 members to be present. Twelve hundred publicans made the journey to Portlaoise and drew up very determined battle lines in their continuing efforts to fight the ban. Attitudes had certainly hardened, and a number of measures were agreed which the members would press to have implemented by their national executive. The most serious measures included refusing to pass on VAT payments, reducing their payments of preliminary tax, and putting forward candidates in the 2004 local elections. Emotions were running very high, epitomised by the declaration by the convenor of the meeting, Tadg O'Sullivan, that 'Our members are not prepared to put their lives, their wives and daughters and female staff at risk.'

Then came a comment from the meeting that was so over-the-top that it drew a very critical reaction from many sources, including national media, and is thought to have swayed public opinion significantly away from concern for the plight of the publicans. 'Who runs this country?' demanded an impassioned delegate from Munster. 'We do,' he declared. With those two simple words, the battle for the hearts and minds of the many impartial observers of the smoking-ban debate was lost.

Micheál Martin recalls the reaction to the comment that he personally received. 'Next day, hundreds of emails and phone calls started flooding into the Department after that comment. People were furious and saying, you can't take that. People were stopping me in the street, saying you can't give in to people like that, who think they are running the country. You have got to keep going.'

An editorial in the *Irish Independent* the next day condemned the publicans: 'They are not entitled to make silly threats like "marching on the Dáil" or daft statements to the effect that they run the country. And least of all are they entitled to talk of withholding taxes. Publicans must hand over the VAT, just as householders must pay bin charges. If they have a case to make, let the government hear it. But they must not back it with threats of civil disobedience. On taxes and smoking alike, they must obey the law.' The *Sunday Times* commented: 'The massed ranks of publicans railing against the smoking ban look like an increasingly desperate mob. Whatever your views, a campaign of civil disobedience is not the way to win the argument.'

Other editorials blasted the publicans and presented delegates to the Portlaoise meeting with a cool reception on their return home. The comments ranged from 'The phoney war is now over' to 'Mr Publican, your days of riding roughshod over us are over' and 'The vintners have seemingly lost the war.' For good measure, the notion of withholding VAT payments drew a very sharp response from Dermott Jewell, chief executive of the Consumers Association of Ireland. Jewell retorted that, if publicans were not about to pass the tax on to the Exchequer, they had no right to charge this tax to their customers and should reduce their prices accordingly. These sentiments summed up the views of the general public to a substantial extent.

The VFI had mobilised a major challenge to the ban and a

strong demonstration of defiance towards it. Then, in one moment of ill-judged emotion, they had lost the plot. As if the fall-out from the Portlaoise meeting had not been enough to make it a bad day at the office for the VFI executive committee, there was more to follow in the coming days.

The focus of the story now moves from the Irish midlands to the Telstra Dome in Melbourne, Australia. In what was considered to be a very ill-timed move by the senior executives of the VFI, at the height of the anti-ban campaign, chief executive Tadg O'Sullivan and chairman Joe Browne departed the scene immediately after the Portlaoise meeting to attend the Rugby World Cup in Australia, as guests of Heineken. Due to the two key VFI representatives' unexpected absence at what was considered to be the most crucial stage in the publicans' struggle, infuriated publicans began to call for their resignation, some branches threatened to resign en masse from the federation, and a motion of no confidence in the two was tabled for the next executive meeting. As it transpired, the motion was not tabled – because, it was reported, the two absentees did not return from Australia in time for the next meeting.

Funny business, lobbying: one day you're up, the next you're down – or 'down under'.

15

MICHEÁL MARTIN'S LOWEST POINT

The rollercoaster of highs and lows did not apply only to the anti-ban side. Minister Martin was very soon to tumble from the high of overcoming some major obstacles to what he himself described as the lowest point for him in the entire process.

First, though, there was good news to digest, as the latest survey of public opinion on the ban was published in early November. The MRBI poll, commissioned by the Office of Tobacco Control, showed that 81 percent of the public were now of the opinion that publicans should be obliged to comply with the new smoking regulations and should implement an outright ban on smoking in their pubs. Support for the ban ranged from 91 percent among members of higher-income groups to 74 percent among lower-income groups; 61 percent of respondents who were smokers said they supported the ban.

Other good news for the pro-ban side came with the release of figures compiled by real-estate auctioneers which showed that the number of pubs sold in Dublin in the first nine months of 2003 was the highest in three years. So, despite the worries about the feared impact of the forthcoming smoking ban, there was no shortage of investors in the on-trade in Dublin; indeed, according to auctioneer Tony Morrisey, trophy pubs continued to fetch premium prices. He instanced the well-known Yacht Pub in Clontarf, which had recently sold for €8 million, and Sinnotts of South King Street, whose leasehold interest went for €3.2 million. Smaller pubs also continued to sell readily, with the CBRE Gunne agency reporting the sale of three Dublin pubs for a combined €6 million.

Another encouraging item was a new EU-funded study, based

on testing ventilation systems, that concluded that even the most powerful of these systems was ineffective in clearing the air in the pub of harmful substances. The research was carried out at the EU Joint Research Centre at ISPRA in Italy. Dr Barry McSweeney of the centre claimed that, to date, no experiments had been conducted in such detail in the US or elsewhere. The tests, which were undertaken independently and, according to Dr McSweeney, were not intended 'to sabotage the vintners', showed that even the most powerful ventilation systems left a chemical residue that was twice the level that would be considered healthy and thirty times the level found in a room where there had been no smoking.

Another comforting development for the minister was an official policy statement from the Irish Congress of Trade Unions (ICTU) fully supporting the ban and decrying the opposition to it as 'weak and pathetic'. David Beggs, ICTU general secretary, proclaimed that the ban was one of the most serious public-health initiatives undertaken for many years and that his organisation, which represented 700,000 workers, was fully behind it.

There was even some modest respite on the question of the possibility that the smoking of herbal cigarettes would pose particular problems for the enforcers of the ban, who might find it difficult to make the distinction between tobacco and herbal cigarettes. This issue was being promoted by the cigarette-machine operators to demonstrate that enforcement would be a major problem. Charlie Sothern, the sales director of Kelkin Natureproducts, the sole distributor of Honeyrose herbal cigarettes in Ireland, resolved the problem. He announced that he had not been contacted by the vending-machine people and that his message for them was that he had no plans to expand his herbal-cigarette business, which was essentially a niche market.

On a lighter note, media interest had turned to a few relevant 'sideshows' to add a new dimension to the smoking-ban story. One of these concerned the Cobh, County Cork, company, Mod-Erector, which was inundated with demand for its special 'smoking shelters'. These could be located in pub car parks or adjacent open spaces and were being seen by many publicans as the ideal way to get around the 'enclosed places' aspect of the ban. Other enterprising pubs were coming up with all kinds of wheezes to keep their customers under cover, but without infringing on the terms of the ban, such as converting old buses and parking them outside their pubs, as was the case with the well-known Johnny Fox's pub in County Wicklow. And to prove that there is no limit to the inventiveness of entreprenuers, a lingerie manufacturer came up with a novel way to help ladies deal with the ban by kicking the nicotine habit. A new-style bra released fragrances that contained a formula that would supposedly discourage tobacco smoking. As they say, necessity is the mother of invention. But on the news front, that was as good as it got in November for Micheál Martin.

This was a particularly busy period for the Attorney General, Rory Brady. In relation to the smoking ban, one of the serious concerns of the government was whether new health-and-safety regulations would be required to empower the personnel who would police the ban. It was not entirely clear if the regulations under which Health and Safety Authority (HSA) inspectors operated were adequately aligned with the new Department of Health regulations. The hope was that 'cross-authorisation' would be permissible, to allow the HSA to dovetail with the Department of Health. A key issue at this stage, however, was the fact that secondary smoke had not yet been designated under existing health and safety legislation as a carcinogen, which meant that it was not yet listed amongst the prohibited substances which were policed by the HSA. The suggestion of re-designating ETS as a carcinogen was not going unchallenged, however, as employers were nervous that this would leave them exposed to retrospective civil actions by employees. In fact, the employers' group IBEC feared a massive wave of compensation claims along the lines of the former, highly controversial 'army deafness' saga.

In order to deal with this possible loophole, the minister of state for labour, Frank Fahey, was in the process of sponsoring new regulations to define ETS as a carcinogen, but discussions were still ongoing, despite the fact that Micheál Martin had already started the countdown to 26 January with the announcement of the Department of Health's smoke-free regulations. The HSA would have to submit new regulations to the EU; by the second week in November, these regulations were still not ready, and under the three-month rule they would be impossible to harmonise with those that had already been submitted. So the question for the Attorney General related to the ability of HSA inspectors to fulfil their duties as soon as the ban came into play, in the absence of special new regulations. The stance of the Department of Health was that the new law could be implemented on the January date, irrespective of this hiccup. Within days, this situation was to change dramatically, however, and this was to turn out to be the most nerve-racking period in the whole saga for Micheál Martin.

*

One of the key issues that was exercising the minds of the Attorney General and the members of the implementation committee was the definition of what constituted a 'workplace'. This proved to be something of a conundrum: following representations from various members of Cabinet, who had concerns in relation to the impact of the ban on their areas of responsibility, and extensive consultation on these issues with the Attorney General, it was announced that five categories of workplace would be exempt from the ban. In effect, exemptions would be granted for smoking to be permitted in hotel rooms, prisons, psychiatric hospitals, nursing homes and hospices, on the basis that these could be regarded as places of residence or dwellings. Prisons already had a specific exemption under health and safety legislation, based on security and good-order considerations, in addition to being regarded as places of residence for prisoners. These exemptions were introduced despite the determination of the minister from the outset that there would be no exemptions, because of the seriousness of health risks to workers in every enclosed place of work. The announcement was taken by many to be the first phase of a climbdown by the health minister.

This change, announced only two days after the Department of Health had reaffirmed that there would be no further delays, led to criticism of the minister for poor planning and opened the door for further attacks from the IHIA and other opponents of the ban –

although, understandably, the IHIA welcomed the exemption granted to hotel bedrooms. Delays would inevitably follow the introduction of these amendments to the recently announced regulations, because the whole process of the three-month EU submission and notification process would have to start all over again. So the new date for the introduction of the ban became 10 February, but even that date was to change again very soon.

In attempting to explain the new exemptions, Minister Martin said that, after exhaustive discussions with the Attorney General and in order to establish a 'firewall' to protect the Irish government's position against every conceivable potential legal challenge to the ban, he was now adopting a 'belt and braces' approach to the legislative standing of the smoking ban. He added that he would be reminding employers in the exempted categories that they had a duty of care to protect their employees, even though smoking would not be prohibited in their workplaces. The exemption only conferred the right not to be penalised, and all employers still had the right to impose the ban if they so decided.

The Prison Officers Association was first to respond, stating that the minister's decision raised serious questions about the health and safety of its members. Then came an angry response from Jack Nash of the hotel-workers union SIPTU, who stated that there was no logic in exposing his workers to a known risk.

Fine Gael health spokesperson Olivia Mitchell articulated concerns about how watertight the new regulations would be from possible legal challenges, considering that, although this issue had been on the political agenda for many months, potential legal problems were apparently being identified at the eleventh hour. She went on to state that the prospect of further loopholes emerging would be a 'lawyer's dream' but conceded that it would be preferable to have a delay of a few weeks rather than years of legal defence of the smoke-free regulations. Needless to say, the tabloid headline writers had a field day, calling the ban a 'SMOKING GUN' and declaring that, if you want to smoke in a workplace, you should 'GO TO JAIL'.

Responding to critics, the minister said that the amendments had been required because of the possibility of locations such as prison cells, hotel rooms, hospices and psychiatric wards being regarded as 'dwellings'. Because of this potential risk, it was necessary to ensure that the new law would be absolutely watertight. 'The devil is in the detail,' he said.

It appeared that the devil was indeed taking a particular interest in Micheál Martin's affairs during the second week in November. A further delay to the submission of the amendments to Brussels was now being caused by the possibility that, in addition to hotels, bedrooms in B&Bs and hostels might have to be exempted as well. This was causing a lot of wrangling amongst the minister's legal advisers, who at this stage were becoming paranoid about leaving any loophole whatsoever open to challenge. The IHIA and VFI were also attempting to exploit the new amendments by asking their legal teams to examine the position of licensed premises, which were also the dwelling places of publicans, thus adding a further complication to Martin's plans.

While uncertainty continued over a number of days, there were indications of unrest at Cabinet level and a clear impression that, after all the controversy over the summer months, the government now simply wanted to get this item done and dusted. Impatience also began to creep into the tone of editorial comment in the national press, with the *Irish Times* for instance calling on the minister to 'stop dithering'.

The dithering was to last for a few more days, until 12 November, when the amendments were finally notified to the Commission. This meant that yet another new three-month standstill period commenced, so the earliest date for the ban was now 16 February. Of course, if any EU member state submitted any reservations to the new measure during that three-month period, the whole process would be delayed yet again.

The Taoiseach had by this stage decided to intervene, and he did so in a manner that was very supportive of the health minister. 'Never in twenty years in politics have I seen a situation where the government had to examine proposed changes in such detail to try to prevent legal challenge,' he declared. Bertie Ahern went on to mount an attack on the opponents of the ban. He said that there appeared to be groups with 'endless money' who were waiting at every turn to mount legal challenges to every detail of the new legislation. In a thinly veiled suggestion that tobacco interests were linked to the proposed attacks on the new law, Ahern said that the attention to detail in drafting the regulations had been undertaken in the clear understanding that the 'most eminent' people were prepared to challenge the ban for financial gain. Although the Taoiseach had not specified any particular group in his attack, the IHIA rose to the bait and declared itself to be independent and funded solely by donations from its members.

Finally, after a very difficult week or so, when much deliberation had taken place and much advice had been sought and obtained from the Attorney General and other legal heavyweights, the amendments were at last agreed by the government. The new amendments included further exempt locations, including the Central Mental Hospital and sleeping accommodation in B&Bs, guesthouses, hostels and student residences.

These amendments were duly submitted to the European Commission, under the EU Transparency Directive, and were described by the Taoiseach as 'rock solid' and capable of ensuring the implementation of the ban three months later. As a further precautionary measure, the Cabinet approved the introduction of new legislation, which would result in tobacco smoke being defined as a carcinogen and obliging employers to protect their workers from this harmful substance. Health-and-safety legislation would be amended accordingly, and the view was that there would be no requirement to submit the new legislation to Brussels, because of the manner in which it had been framed in consultation with the Attorney General.

So the final administrative obstacles had been removed, and it was all systems go to begin the ban in February 2004. If only life was that simple.

*

Further delays to the implementation date were feared as a result of the possibility of 'opinions' on the proposed changes to the Irish law being expressed at EU level and subsequently leading to a winding back of the clock and a restarting of the three-month 'standstill' period. Since the 1980s, EU 'transparency' obligations had been imposed on governments, under which it became necessary to inform the European Commission about the introduction of any regulations that might have implications for competition law. The objective of this approach is to afford member states the opportunity to express their views or detailed opinions on a particular matter. A government is not obliged to change proposed measures because of comments made by other member states, but where the tobacco industry is concerned, not to make changes tends to be regarded as a high-risk strategy, as it could lead to defendants in future civil cases calling on EU governments in their defence. There were indications that Germany and Austria might have problems with Ireland's proposed regulations on smoking in the workplace, so it was by no means certain that complications would not arise during this standstill period.

The minister admits that this period was the 'low point' in the whole process for him. 'I think we took a hell of a lot of stick and brickbats, which I had to sit and absorb,' he recalls. 'I remember reading every negative article at the time we announced the whole business about exemptions and delays. I got a sense that the whole world wanted to beat me up badly over this, but that was fine because I knew the "promised land" was nigh. We were going to get there and that was the time that resolve was needed. It could be argued that we were over-cautious in relation to detail, but it was time for caution, because we knew the world of tobacco is bedevilled with legal issues. So we took the hit, knowing we would stand steadfast and knowing it was about procedural issues rather than anything blocking the ban. We were confident about that, but we certainly did have to absorb a lot of flak.'

It was November, and there was still a while to go. The date for introducing the ban had still not been finally fixed.

16

FINALLY, THE END IS IN SIGHT

With the regulations and amendments finally drafted and submitted, heading into December the media spotlight moved onto other seasonal issues, and the brunt of the negative feelings against the government switched to disappointment with certain aspects of the Exchequer Budget and away from the bureaucratic hurdles facing the smoking ban. This gave the pro-ban side a short breather following a frustrating and sometimes demoralising few weeks in November. They now needed to step up a gear with the smoking-cessation programme, however, in anticipation of people's traditional New Year resolutions to quit smoking. A wide-ranging campaign was on the drawing board at the Health Promotion Unit (HPU) to maximise the impact of what was likely to be the biggest-ever incentive for Irish smokers to quit. In conjunction with the Irish Cancer Society and the health boards, the HPU was planning to direct as many cigarette quitters as possible to the Quitline and other counselling services.

While this very critical phase of the Department of Health's action plan was being put in place, the IHIA was continuing to obstruct the new legislation at every turn. The IHIA contended that their legal advisers had spotted a loophole in an EU directive and that this loophole could set the Irish ban back by at least another year. This was in relation to the possibility of a Europe-wide anti-smoking initiative being considered by the European Commission; such an initiative could technically undermine the Irish ban, on grounds of duplication.

Much to the relief of the Irish legislators, the European Commission moved to confirm that the procedures followed by the

Irish government were in order and would not be in conflict with any current plans at EU level and therefore would not be hindered by the technicalities as specified by the IHIA lawyers.

The next ploy used by the IHIA was to focus on their claim that the ban would be unenforceable, due to the limited number of inspectors and the continuing possible uncertainty about the empowerment of health-and-safety inspectors to operate within the new regulations. The alliance was also attempting to highlight problems resulting from the fact that implementation guidelines for publicans and hoteliers still had not been circulated, with less than two months to go before the launch of the ban.

The main thrust of the guidelines was duly flagged in a leaked story at the end of December, while the final texts were still being drafted by the implementation committee. They would advise publicans to take a hard line with offenders – to refuse to serve them drinks and, if necessary, to call the police. They also required the use of permanent 'no smoking' signs and posters to be prominently displayed in all enclosed workplaces, and the removal of ashtrays.

*

One of the most significant developments in the entire campaign came about during the Christmas 2003 period. Following a meeting of its sixty-member national executive council, the VFI announced at the end of December that it was advising its nationwide membership to implement the ban. In what amounted to a radical change of attitude from its previous hostile, confrontational approach over the preceding eleven months, the VFI said that it would continue to keep open the option of a legal challenge but would not suggest that its members break the law when it came into force. The VFI continued to express serious concerns about enforcement of the ban, however. Importantly, though, there was a clear indication that publicans were now facing up to the fact that the ban was coming and that they had better make the best preparations for it they could in the time remaining.

Meanwhile, the phones were starting to hop as smokers inundated the Quitline with calls for assistance. In the first week of

January, some 2,000 calls were dealt with. Norma Cronin of the Irish Cancer Society monitored the nature of the calls and found that the impending smoking ban was the key motivating factor in encouraging many people to grasp the nettle and try to give up smoking. Never before had smoking been on the everyday agenda of so many people for such a long period of time; would-be quitters were comforted by the fact that they were part of a massive collective cessation movement.

According to Chris Fitzgerald of the Health Promotion Unit, since the previous November a total of 8,500 calls to the Quitline had been dealt with, the callers being equally split between men and women and predominantly coming from the twenty-one-to-forty-year age cohort. He also commented that available research indicated that 70 percent of smokers wanted to quit.

'Any news on the date?' That was the question regularly asked of Caitriona Meehan in the minister's office and of Rachel Sherry, the PR consultant, on her daily communications assignments. As January dragged on, the date still had to be fixed, and there was a certain anxiety about the possibility of a spanner being thrown in the works by Brussels.

Behind the scenes, the communications-strategy teams in the department and at the Grayling agency were busy weighing up the pros and cons of the various options for dates, subject always to developments at European Commission level. Everybody, including the minister, was anxious to go for the earliest possible date.

To go for the first available date of 16 February was still attractive. It would be good to put an end to the continuous media speculation. There was still the possibility of a further setback, however, if any objections came from Europe. Would it be too much of a risk? There was certainly no appetite for announcing any further date changes. Everybody was agreed that the next announcement should be the final one.

If objections did arise, there would be a further ninety-day postponement. The start date for the ban would be pushed out to the end of May – the eve of the local elections. Ouch!

A further consideration was that many in the health-alliance team were concerned about some other practical issues, such as ensuring that sufficient time would be allowed between the time of the announcement of the new ban date and the ban coming into effect. There would still be many pieces of the information jigsaw to be fitted into place, especially for the guidance of employers, in that defined period.

It was something of a catch-22 situation: realistically, no announcement could be made before the Brussels deadline of 16 February had passed. The team would simply have to live with the ongoing media impatience and pressure for information. Finally, on 18 February, the air was cleared.

*

A huge collective sigh of relief emanated from the pro-ban team when word came through that no objections had arisen at EU level. It was all systems go. Early on the morning of 18 February, Micheál Martin walked into a crowded room at the Mont Clare Hotel, where Chris Fitzgerald was chairing what was to be the final meeting of the steering committee. They had not yet heard the news, so when the minister announced to the group that he was heading immediately to a media briefing in Government Buildings, to announce publicly that Ireland would go smoke-free in workplaces from Monday 29 March, there was a feeling of elation. He thanked all concerned for the important role that had been played by the alliance of health professionals and social partners, who had cooperated to an exemplary extent in helping to pave the way for the introduction of a historic new Irish law.

Then on to Government Buildings, where once again a packed media gathering, which included international reporters and TV crews, was informed that the final countdown to the ban had begun. The news was greeted with a sense of relief by all those present and was reported very positively. Thirteen months after Minister Martin's original announcement of plans for the ban, all the elements were finally in place for the implementation of one of the most widely publicised pieces of Irish legislation ever.

There were still to be a number of interesting developments, though. On 4 March, it was announced by the Licensed Vintners Association that its Dublin publican members would enforce the ban. Days before the ban came into effect, all the ministerial regulations and amendments and the whole EU approval process was superseded by the Irish government's separate decision to update primary legislation giving the Dáil and Senate absolute powers in relation to banning smoking in Irish places of work. This did not need any further referral to Brussels and effectively copper-fastened the ban against legal challenge. The government was not taking any chances on this one. It was also ensuring that any potential future changes would not be possible without full parliamentary approval.

There was also something of an ironic twist in the fact that, on the evening of 28 March, the day before the launch of the smoking ban, Chris Fitzgerald of the Health Promotion Unit (HPU) represented Minister Martin at an event that was sponsored by the HPU that was also attended by Tom Beegan of the Health and Safety Authority, whose inspectors would police the ban from the following day. Both had been amongst the prime players in the lengthy preparation process leading to the introduction of the ban; at the conclusion of the event they moved through the smoked-filled lobby of the City West Hotel – a hotel lobby that would be smoke-filled for the very last time.

The irony comes in the fact that the Irish Masters Snooker Championship was being held at City West: the event was being sponsored as a vehicle to promote anti-smoking messages to a specific target audience. Some years previously, the very same championship had been the last sporting event to be sponsored in Ireland by a tobacco firm – by the cigarette company Benson and Hedges.

The smoking ban came into effect the next morning, 29 March. Within the first week, there was 97 percent compliance with the ban in the hospitality industry. By the way, on 29 March all ashtrays had been removed from Jackie Healy-Rae's pub in Kerry.

EPILOGUE

Reflecting on the launch day of the smoking ban, the Minister of Health was quick to admit that, despite his earlier determination to bring in the ban at the beginning of 2004, there was a distinct advantage in the postponement of the introduction of the ban from the original cold, damp launch date of 1 January to the considerably warmer and brighter start date at the end of March. Indeed, the March date was a lot closer to the late-spring or early-summer date that many of his advisers had strongly recommended from the outset. Reflecting on the launch, Minister Martin said: 'In hindsight, we were blessed that, due to having to notify Brussels, the ban was put back to March. It would have been more difficult in January and it was especially in our favour that it turned out to be a fine sunny day when the ban came into force at the end of March.' Of course, the ironic twist is that, if the government had opted in the first instance to introduce the ban through implementing primary legislation (by amending the 2002 Tobacco Act), as eventually transpired, instead of taking the route of detailed EU notification, the January date probably would have stood. It would appear that the reason the 2002 Act was not notified to Brussels in the first instance was that so many of its provisions had been challenged by the tobacco industry. These provisions were subsequently ratified and were incorporated into the 2004 Act, under which the ban was ultimately introduced.

Another very interested individual who welcomed the postponement until March but still experienced a degree of trepidation as the launch date approached was John Douglas of the bar-staff trade union Mandate. 'From the feedback we were getting from publicans, especially in the Dublin city and suburban pubs,' he said, 'it was clear that right up to the eve of the ban's introduction they still believed it would not happen. On the morning the ban came into effect, I drove down Parnell Street in Dublin, past some of the "early houses", not knowing what to expect. I was very pleasantly surprised to see clusters of burly men, having finished their early-morning shifts at the markets or the docks, standing outside the pubs, quietly smoking their cigarettes. I never thought I would see the day.'

Douglas did not receive a single report of any problems or difficulties encountered by his union members as a result of the ban being introduced. 'It was as if everybody had been conditioned well in advance, as a result of all the discussion and media debate, and they were well prepared for it when 29 March finally arrived,' he commented. 'Our bar-staff members told us that they never realised how unpleasant the environment had been in their workplaces prior to the ban. They would arrive home after a day in the pub, feeling unclean and smelling of tobacco smoke. Now they feel so much better and would never go back to the old days.'

*

Very high levels of compliance with the ban were immediately apparent to the environmental-health and health-and-safety officers during the first few days of the ban. The very high levels of public and trade support for the ban in the period immediately after 29 March were also confirmed by the media. As had been widely anticipated, reporters were despatched far and wide to seek out publicans and customers who were prepared to defy the ban. They found very few candidates and concluded that the ban was a resounding success. The instances of non-observance of the ban, even in pubs in the toughest neighbourhoods, proved to be so limited as to be considered inconsequential. In fact, the two enforcement agencies reported soon after the launch of the ban that 97 percent compliance had been achieved in all workplaces, including pubs.

*

Within a week of the ban being enforced, the first high-profile report of defiance of the ban appeared in the media – somewhat appropriately on 1 April. The fact that the incident occurred in a parliamentary setting added a great deal of spice to the affair, for national and even international media.

John Deasy, the Fine Gael Party spokesperson for justice and Waterford TD, openly defied the ban by smoking three cigarettes in the Dáil bar. Despite protests from the bar staff, he refused to put out the cigarettes, claiming that he was entitled to have access to an outdoor smoking facility in close proximity to the bar but that none had been provided.

This was not the first time that Deputy Deasy had courted controversy and had wound up being featured in national and local media as something of a rebel. He had previously publicly criticised the Fine Gael Party leadership and had declined to vote with his own party on the contentious Immigration Bill. On this latest occasion, however, he paid a significant price for his high-profile defiance. He was promptly criticised for his undisciplined and irresponsible stance and was sacked from the Shadow Cabinet by Fine Gael leader Enda Kenny, who found the whole affair intolerable, especially as it ran completely counter to the party's stance on the smoking ban and represented a breach of the new law.

Another high-profile demonstration of defiance of the ban was to hit the headlines and broadcast-news bulletins during the summer, when the Fibber Magee pub in Eyre Square, Galway, announced that it would discontinue observing the ban due to a dramatic fall-off in business. The pub claimed that there had been a 30 percent fall in bar sales since the introduction of the ban.

In one of the most widely covered news stories in July, the owners of the pub announced in a news interview on TV that they would be putting out ashtrays and inviting drinkers who wished to smoke to come to their pub and do so at will. Environmental-health officers immediately issued warnings to the pub owners, but the pub nonetheless proceeded with its plan and ignored the ban, and many locals and tourists flocked to the pub to see what would happen as a result of such a highly publicised rebellion.

With a number of pub owners from other parts of the country

coming along to see the action and announcing that they would return to their premises and ignore the ban too, a serious challenge was beginning to arise in relation to the enforceability of the ban. The Western Health Board moved very smartly to show that it was not about to be intimidated by rebellious publicans and was determined to prosecute any pubs that would not cooperate with the environmental-health officers in observing the ban. Such prosecutions could have serious consequences for pubs that were seeking to renew their licences. The prospect of conviction and the serious consequences thereafter served to bring the publicans to their senses, and the skirmish of defiance proved to be short-lived; in fact, it lasted only two days.

Some months later, the owners of the Fibber Magee pub appeared in court. Despite pleading that their action was 'a rush of blood to the head' in response to falling profits, they were prosecuted and fined €9,400. In addition, the owners of the pub were fined €200 each for smoking on their own premises and were advised that a number of their customers who had been found smoking would also be pursued.

The claim about falling profits had obviously cut no ice with Judge Mary Fahy, who handed down the sentences. Firstly, this claim was irrelevant to the court, and secondly, the judge was no doubt just as aware as everybody else in Galway that, with major roadworks disrupting all businesses on Eyre Square throughout the period and students being on holidays in July – Fibber Magee's was a well-known student haunt – the only purpose of the smoking-ban stunt had been to try to increase business through notoriety.

The story did not end there. At a further hearing of Galway District Court the following November, the Western Health Board objected to the transfer of the licence for the Fibber Magee pub. In response to the objection, one of the publicans who had been prosecuted for defying the ban went into the witness box to give a formal undertaking that he would abide by the new legislation. He also called on all publicans and members of the public to comply with the ban. In addition, he agreed to pay the legal costs of the Western Health Board for the court case. Judge Mary Fahy duly granted the licence certificate.

Within two weeks of the ban coming into effect, the Irish College of General Practitioners announced the results of a survey showing that 28 percent of smokers were planning to quit. A further observation on the immediate impact of the ban came from Valerie Coghlan of ASH Ireland, to the effect that many heavy smokers had reduced their tobacco consumption as a result of their smoking having been curtailed in workplaces and pubs. ASH noted that many forty-a-day smokers had cut back to twenty or twenty-five cigarettes a day.

Six months into the ban, Norma Cronin, the Irish Cancer Society's smoking-cessation expert, announced that a study conducted by the research company Behaviour and Attitudes indicated that, by September, 7,000 smokers who had contacted the National Smokers Quitline had given up cigarettes. A further 48 percent of smokers who had contacted the Quitline – 9,500 people – had not given up smoking but had cut back.

Of the 19,800 smokers who called for assistance, 64 percent had been smoking for more than fifteen years. Twenty-four percent of callers were under thirty years old, and 43 percent were over forty. Cronin proclaimed that the new legislation, as well as being effective in protecting the health of workers in previously smoke-filled environments, had 'been instrumental in providing a positive motivating influence in pushing smokers to make that difficult but extremely worthwhile decision to quit smoking cigarettes.'

Reports from the pharmaceutical companies that produce nicotine-replacement therapy (NRT) products confirmed the news that thousands of former smokers were now taking action to stop. Demand for NRT products increased dramatically in the period when the smoking ban was launched, levelling off to show a 6 percent increase for 2004 as a whole, compared to 2003. Sales of the fastest-growing NRT brand, NiQuitin CQ, increased by a formidable 19 percent in 2004.

In tandem with the drop in the number of smokers, there was a 7.5 percent fall in cigarette sales in the first six months of 2004. This meant that 260 million fewer cigarettes were sold in Ireland during the period when the workplace smoking ban was entering its first phase. By November, some eight months into the ban, the slump in cigarette sales had gathered further momentum. Exchequer estimates on the eve of Budget 2005 showed that a decline of 17.6 percent in tobacco excise revenue had occurred in 2004, indicating a fall in cigarette consumption of between 700 million and 800 million in the first full year after the introduction of the ban. This would result in a drop in Exchequer tax take of €128 million. Fears that this might influence the newly appointed Minister for Finance, Brian Cowen – Micheál Martin's predecessor as health minister – to revisit the effects of the smoking ban were dispelled when Cowen said that the ban was achieving important health benefits. Although he did not rule out further increases in the tax on cigarettes, no such hikes were introduced in Budget 2005.

*

One of the more interesting social phenomena which was latched on to by the Irish media was the new 'meeting point' generated by smokers moving outside to the various outdoor smoking facilities created by progressive publicans. It was even claimed that romance frequently blossomed amongst smokers in the open air.

A survey conducted by the magazine *Hospitality Ireland* concluded that a new phenomenon known as 'smirting' had emerged as a result of the smoking ban. Smirting was the term – created in customary Irish fashion whenever a topic catches the public imagination – to describe pub smokers who flirt. Indeed, one in four adults surveyed claimed to have been attracted to a fellow smoking-ban observer and to have had as much, if not more, fun outside the pub than they had inside the same establishment before the ban came into force.

Another, more serious, survey conducted by the Department of Health in August 2004 showed that public support for the smoking ban was holding firm. Eighty-two percent of adults supported the Smoke-Free at Work legislation; 90 percent regarded the ban as beneficial to workers.

*

Publicans continued at every opportunity to attempt to raise objections to the ban and to highlight problems with it, especially in relation to falling turnover associated with the ban. Significantly, however, the Licensed Vintners Association mounted an unprecedented publicity campaign – interpreted by many as something of a U-turn – promoting the improved smoke-free environment in their members' pubs. The campaign was based on the theme 'The atmosphere's got even better' – referring to the pubs' new smoke-free environment.

Repeated references to the negative impact on business attributed to the smoking ban never succeeded in gaining credibility with the Irish media. Invariably, such claims came to be treated by commentators as merely one factor in a multifaceted scenario, with diminishing pub business just as likely to be due to what were perceived to be exorbitant prices, as well as demographic and economic factors. The pressure on young people to extend themselves financially in order to meet large mortgage repayments, as house prices continued to soar and with those houses located further and further from places of work, came to be recognised as one of the major influences in switching drink sales from pubs to off-licences. The costs of baby-sitting and taxis were also cited as further factors motivating young couples to stay at home, entertain at home – and perhaps also smoke at home.

While publicans, nevertheless, continued to emphasise the smoking ban as a primary obstacle for them, there was a change in emphasis when, in November 2004, as part of their pre-Budget lobbying campaign, the Licensed Vintners Association began to focus more on the impact of excise duties on price increases. The vintners argued that alcohol taxes in particular should not be increased, challenging a call from the government's Task Force on Alcohol to introduce higher taxes as a deterrent to alcohol abuse. The vintners pointed to the detrimental impact on their business of the hike in the tax on spirits in the 2003 Budget. This, it was claimed, had caused a 20 percent decline in pub sales.

An economic study was commissioned by the Drinks Industry

Group to highlight further to the Minister for Finance the problems faced by the publicans. Interestingly, media coverage of the study, which was conducted by Tony Foley of Dublin City University, emphasised that, taking the large differential between prices in pubs and those in off-licences into account, customers were now making almost half (by value) of their alcohol purchases from the take-home trade. A further point which attracted media attention was that, despite falling sales volumes, publicans' revenues were holding up; this suggested that, when the pub trade came under pressure, the automatic response from publicans seemed to be to increase prices.

When details of Budget 2005 were announced, it transpired that no tax increases were applied to either cigarettes or alcohol. This was the first time in twenty-six years that the 'old reliables' had been left untouched in an Irish Budget. This approach drew criticism from Luke Clancy of ASH Ireland, who proclaimed that a further opportunity to deter smokers had been lost. There was also a good deal of speculation that the publicans lobby may have retrieved some of its influence over politicians and effectively sold the proposition that they had already been so badly hit by declining business that they should be cushioned against any further impact caused by tax-fuelled price increases. Some speculation went so far as to suggest that, in return for their cooperation with the smoking ban, the publicans had been promised benign treatment in Exchequer Budgets for some time to come.

*

Media interest in the ban continued at a very high level right up to the day it was launched. There was a particularly large media presence, from both Ireland and overseas, at a special celebration breakfast at Bewley's Café in Dublin on 29 March. The occasion was hosted by ASH Ireland, whose chairman, Professor Luke Clancy, recalls: 'I felt that something really important was happening. The smoking culture was changing. The workers of Ireland had accepted that second-hand smoke was bad. Minister Martin and the government had faced down the publicans and the tobacco industry

155

had hardly dared to show its face during the debate. The world came to our breakfast in Bewley's and Ireland was not afraid to be first and seemed to be proud of the role we were playing in tobacco control. I felt we were on our way!'

Two days after the introduction of the ban, the national media had moved on to other news stories. The smoking ban continued to attract a small number of 'Letters to the Editor', but as far as editorial comment went, the ban story was history. Nonetheless, over the preceding fifteen months it had certainly delivered a diverse and exciting chapter in Irish media debate.

Looking back over the intense media battle relating to the ban, some interesting observations arise. Tactically, the anti-ban campaigners had scored very well in capturing a significant share of media attention in July and early August, maximising the opportunities presented by the 'silly season'. While the protestors generated a high volume of print and broadcast coverage, however, in terms of positive impact the supporters of the ban had scored a decisive victory, according to a comprehensive media analysis conducted by MediaMarket, an independent agency that carries out media monitoring and evaluation assignments. The agency undertook a detailed analysis of the media coverage of the eighteen-month-long smoking-ban story, for presentation to the Irish PR industry, as a means of demonstrating the nature and breadth of the services it offered.

The main findings of the survey of 5,996 print-media stories (broadcast media were not covered), representing coverage of 1.8 million square centimetres of media space, showed that the pro-ban lobby had secured 65 percent of 'share of voice', based on the coverage over the full period. Significantly, the ban supporters had achieved 70 percent of the 'positive-influence' element of the overall coverage. (This form of analysis breaks down the editorial influence the coverage delivers, under the headings 'positive', 'neutral' and 'negative'. The coverage generated by the pro-ban side was found to be predominantly positive in its influence.)

Among the pro-ban players, Minister Micheál Martin predictably emerged as the dominant contributor and spokesperson, achieving 34.5 percent of share of voice, followed by the medical profession, with 10.5 percent, the Office of Tobacco Control, with

9 percent, and NGOs such as ASH Ireland, which achieved 3 percent – all of which coverage was positive.

On the anti-ban side, the survey recorded that the Vintners Federation of Ireland achieved 16 percent of share of voice – significantly more than the specially formed Irish Hospitality Industry Alliance, which scored 8 percent.

The predominant messages that were identified from the ban supporters' campaign were:

§ countering the arguments from the opponents of the ban
§ highlighting the benefits of quitting cigarette smoking
§ encouraging support for the ban
§ explaining the need to protect workers in smoky environments.

The reasons for the success of the pro-ban campaign, as identified by the survey, were primarily that their messages were consistent and clear-cut, were based on facts and expert knowledge, and were focused on health issues such as saving lives and protecting workers who were vulnerable. This would have been seen as distinctly different from the approach adopted by the objectors, whose main thrust was business- rather than health-orientated.

Following a detailed analysis by the author of all Irish media over a sixteen-month period, it is estimated that the smoking-ban story occupied in excess of 20 million words, equivalent to some 10,000 pages of newsprint and 2,000 hours of national and local broadcast time.

*

In the months following the start of the Irish ban, there was widespread interest in the effectiveness of the ban from many tobacco-control interests worldwide, particularly in countries where similar initiatives were being planned or contemplated. In May, Minister Martin visited Norway, the next European country after Ireland to ban workplace smoking. Delegations from Sweden and the UK, including from Scotland, Liverpool and Plymouth, visited Ireland to discuss all aspects of implementation of the ban, and requests for information were received from places as far-flung as Japan, Iceland, New Zealand and India. Sweden was the next country scheduled to follow Norway with a total ban, on 1 July 2005. In Canada, since the Irish ban was introduced, total bans have come into force in the provinces of Manitoba, Nunavut, New Brunswick and Saskatchewan.

The governments of the Australian Capital Territory, New South Wales and the state of South Australia have announced plans to introduce legislation to ban smoking in all restaurants and bars. A total ban is planned by 2006 for Western Australia, where, apart from workplaces, smoking is currently partially banned in night-clubs. In Asia, Bhutan is aiming not only to introduce smoke-free legislation but also to eliminate all tobacco use from the country. Singapore is moving to introduce a ban in all bars, in addition to its existing ban on smoking in most workplaces. In short, the world-wide movement to ban smoking in workplaces is continuously gaining momentum.

Meanwhile, although there are still 1.3 billion smokers world-wide, as more and more controls on smoking are introduced across the globe, the tobacco industry has been moving to exploit market opportunities in developing countries. The World Health Organisation (WHO) estimates that 64 percent of the world's smokers now live in developing countries. In China alone, there are 300 million smokers, and tobacco consumption is continuously increasing in that country, assisted by a favourable tax regime, due no doubt to China being the largest producer of tobacco in the world.

The WHO estimates that tobacco smoke causes 4.9 million deaths annually; unless action is taken, this figure is expected to double in twenty years. The agency closely observed developments in the lead-up to implementation of the Irish ban and was highly complimentary of the role played by the Irish health minister and Irish tobacco-control agencies. The agency has stated its wish that other countries follow Ireland's example on this issue.

*

Almost a full year into the ban, one of the most vocal objectors to it, Tadg O'Sullivan of the Vintners Federation of Ireland, was asked if he had moderated his views or wished to comment on the battle that had been waged against the ban. He responded: 'We realised from the very beginning that some form of ban was inevitable. Indeed, my own campaign to head it off began in 1996. A ban on smoking in certain circumstances would always have been acceptable. From the very beginning, we wanted the right or opportunity to provide for our smoking customers as well as protecting both the staff and the non-smoking customer.

'It haunts me the way in which the medical so-called experts manipulated the scientific evidence and were so dishonest in their zeal to scare the public. A zealot on a mission is a dangerous animal and will invariably write bad law if given the chance. Blatant lies were told by the pro-ban lobby and they raised genuine fears amongst people who should have been advised, not misled.

'Because of the distortion of medical and scientific evidence and the way in which the general public was misled and terrified, sometimes with the connivance or at least the passive observance of the media, it was clear from early on that the battle would not be won in the media. We had assurances from a large number of individual politicians that they would favour the kind of compromise that we were proposing. Unfortunately, those politicians who were supportive were outflanked by the Cabinet or their front benches.

'Finally, I am still sufficiently optimistic to believe that reason will prevail and that those engaged in the hospitality industry will be allowed to provide for their customers who smoke in some form of comfort rather than them being herded like criminals into various forms of shelter. When it becomes apparent just how serious the loss of employment has actually been, once it becomes apparent the impact this ban has had on our tourist industry and on the Exchequer, common sense will come to the surface and the minor compromise that is required can then be achieved.'

*

It is only fitting that the man who started this whole story, the then Irish Minister for Health, Micheál Martin TD, should conclude this account of one of the most significant health initiatives ever launched in Ireland, which subsequently also served as a benchmark for many other countries and states where the fight against tobacco smoking goes on. On the last day on which Minister Martin served in the government's health portfolio before moving on to new challenges as Minister for Enterprise, Trade and Employment, he had this to say in relation to the smoking ban: 'We should acknowledge that the smooth commencement of the ban on 29 March 2004 was not by accident. The public came on board, and so also did the pubs. The Taoiseach, Bertie Ahern, had been talking to publicans, especially in Dublin, and in the last month before the ban came into effect the publicans came around and said that they were not advising anybody to abuse the law. They were going to abide by the law. This served to settle things down. They were not saying that they agreed with the ban, but they had made their points, they understood that the ban was going ahead, and they were going to stay within the law.

'I was never unduly pressurised by the resistance from the publican side – indeed, I was invigorated by the debate. My determination was based on my belief that people have a right to work in a healthy environment. I am convinced that somebody was stoking the debate for all the wrong reasons, and certainly not on the basis that this was a public-health issue.

'The lives and health of members of the public were at stake. That's why we took the decision to ban smoking in the workplace in Ireland.'